Guide to Webcams

Guide to Webcams

By Jason Byrne
& John Breeden

PROMPT®
PUBLICATIONS

PROMPT© Publications is an imprint of Sams Technical Publishing, 5436 W. 78th St., Indianapolis, IN 46268.

International Standard Book Number: 0-7906-1220-8

Library of Congress Card Number: 00-107163

Acquisitions Editor: Alice J. Tripp
Editor: Will Gurdian
Assistant Editor: Kim Heusel
Typesetting: Will Gurdian
Proofreader: Kim Heusel
Cover Design: Christy Pierce
Graphics Conversion: Christy Pierce
Illustrations and other materials: Courtesy the authors

PRINTED IN THE UNITED STATES OF AMERICA

9 8 7 6 5 4 3 2 1

DEDICATIONS

For Katherine, Larry, and Leslie, and everybody else who has always believed in me. My accomplishments are greater and mean more to me because of you.

Jason Byrne

For Karen. Thanks for believing and for waiting all those long nights while I was busy peering at the world through the eyes of a webcam. Your eyes are far more beautiful.

John Breeden

Contents

PART 3: Getting Started With Applications

ABOUT THE AUTHORS

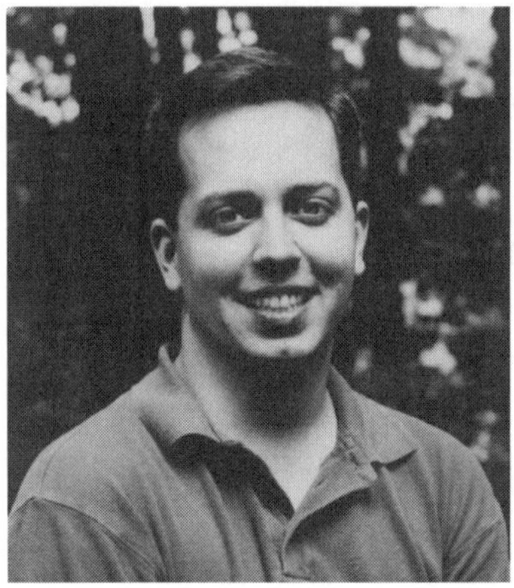

Jason Byrne started using computers at the age of 10 and got his first paying computer job by the time he was 12 years old. Having started with such classic platforms as the Texas Instrument TI 99/4A and the TRS-80, he graduated to the first IBM PC model. He began to learn programming, mainly Pascal, and delved into the mysteries of dBASE.

While pleased with the intricacies of Pascal, he felt limited by the lack of any kind of real graphic support. In 1987, upon getting a summer job at one of the first Egghead Software stores on the East Coast, he quickly began to learn the Macintosh and began using MacPaint.

After spending hours and hours programming, developing digital art, and generally submersing himself in computers, Byrne left for college, where he promptly chose a major in political science. Two years later, a summer internship at the U.S. Environmental Protection Agency turned into a full-time job as a part-time network administrator and a full-time system support technician. Also added to his duties was the task of developing graphical presentations.

In the early '90s, Byrne left his government job and went into business for himself, free-lancing as both a graphic designer and a technical support specialist. Some years later, tiring of working for himself, he joined Cable and Wireless as a system administrator, replacing an outside contractor. A short six months later, he and his entire department were outsourced to an outside contractor.

Byrne became involved in a number of other pursuits, especially politics, designing literature for campaigns. But he was, of course, still fixing, building, and/or using computers. In 1996, he joined the staff of *Government Computer News*, a trade publication, as an associate editor for reviews, his first paid writing job. There, Byrne was to evaluate and write reviews of hundreds of products, as well as write regular columns and many feature articles.

During his tenure at *Government Computer News*, Byrne expanded his technical experience across many areas and stretched his writing skills. Having started off simply as a network product reviewer, by the time Byrne left, he had his own column in every issue of *GCN* and its sister publication *GCN State and Local*. He also had a column in the quarterly magazine *Government Windows NT* and wrote many feature articles and buyers' guides for *GCN*, *GCN State and Local*, *GCN Shopper*, and *Government Windows NT*. *The Washington Post*, which owns *GCN*, has rerun a number of Byrne's reviews in its Tech-Thursday section.

In 1998, he was promoted to senior editor. In the summer of 1999, Byrne appeared as a panelist on the nationally syndicated *Diane Rehm Show* on National Public Radio to discuss wearable computers. In late 1999, Byrne left *Government Computer News* and began his free-lance career again, this time as a writer. *Guide to Webcams* is his first book.

Byrne will be 30 years old this year, and is married to a most forgiving computer widow, Katherine. He can be reached at jbyrne@pressroom.com.

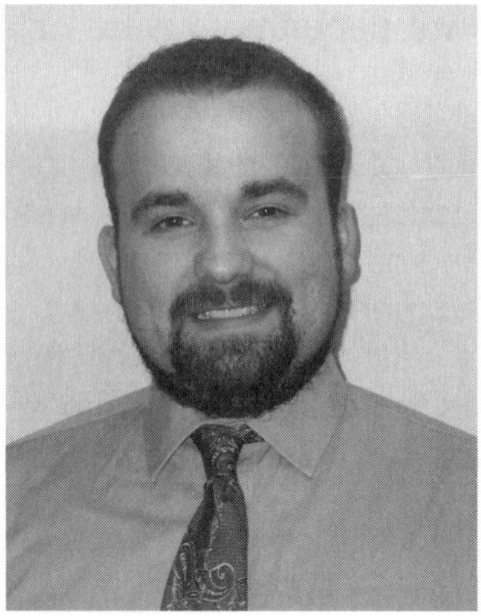

John Breeden II is the associate editor for technology reviews for *Government Computer News*, a technical magazine owned by *The Washington Post*. He has held this position in the *GCN* Lab for three years and is currently in charge of web-based reviews, high-end software packages, and the expanding world of notebook computers. He also writes for *The Washington Post* Friday newspaper, doing reviews for the Fast Forward section. His weekly reviews appear both in *GCN* and in the *Post* business section on Thursdays.

Prior to joining the *GCN* Lab, he worked on the news desk of *GCN*, where he covered technology issues in the military, civilian agencies, and the U.S. Congress.

He is currently the vice president of Noble Order Press Enterprises, a technology start-up firm he founded with his business partner Nate Wooley. NOPE specializes in targeted e-mail newsletters that are produced for high-tech firms looking to increase consumer awareness and confidence in their products. Their flagship publication, *Game Industry News*, is a weekly trade publication covering the computer-games industry. Breeden serves as the editor-in-chief of that publication, managing an extensive free-lance staff of writers and reviewers as well as a small permanent staff of editors. *Game Industry News* can be accessed on the Web at www.gameindustry.com.

Breeden has been a featured writer for several "ask the computer guy" columns in magazines such as *Up Front!* in

New Mexico and *Who Cares?* in Washington, D.C. He has also been a speaker on television and radio programs. Most recently, he appeared on Voice of America European Television and spoke to Eastern European entrepreneurs about how to set up and maintain a computer-games development company—and the best ways to break into the lucrative North American market.

His first venture into writing about technology was started when he founded a weekly column "On the Chip Side," which grew to a circulation of more than 1 million and was published in newspapers in several states.

Breeden holds a B.A. in mass communications from Frostburg State University. An avid e-mail checker, he can be reached at breeden@pressroom.com or editor@gameindustry.com.

PART 1:
A Little Background

What is a Webcam, Anyway?

Chapter 1

This book is about digital video cameras, also called webcams, or cams for short. A few years ago, nobody had even heard of them, and now millions of people worldwide are using them. My aim in writing this book not only is to explain how to use webcams, but also to provide you with background information about the cam community, to familiarize you with webcam terminology, and to show you how to get the most out of your own cam.

A few years ago, nobody had even heard of them, and now millions of people worldwide are using them.

I'll also tell you what kind of computer you'll need, describe the ins and outs of software packages that interact with webcams, and generally prepare you as much as possible to get a positive experience from this hot new technology. Hopefully, with some of the examples I'll tell you about, I'll also inspire you to try new things and come up with your own applications.

If you're reading this book, then most likely you're either thinking about buying a cam or you already have one. Due to the broad range of skills and experience out there, I've set up this book so that it can either be read from start to finish or simply be used as a reference guide. You can skip over the areas you're familiar with and concentrate on the chapters that contain new information.

While the common complaint about technology has been that it isolates people, many new technologies such as webcams are actually bringing people closer together in a number of intriguing ways.

What it's all about

The computing world has seen many watershed events in its short history. While the underpinnings that created the entire industry cover a broad span of time, 99 percent of the advancements in computers have happened in less than the average human life span. Most of what is in your computer today wasn't even invented 25 years ago. And for each quantum leap in computing power, for each new technology, the impact of computers on your life has increased. Now, with the expansion in technologies surrounding the Internet, including webcams, you have more power than ever to tailor your interactions with computers to suit you—rather than the other way around.

While the common complaint about technology has been that it isolates people, many new technologies such as webcams are actually bringing people closer together in a number of intriguing ways. Not only are you better able to keep in touch with those you already know, but you also can meet new people from around the corner, in the next state, or halfway across the world. You can sit in front of a computer in New York and watch the sunset in Key West, Fla. You can talk face-to-face with distant cousins who live on distant continents.

Part of the American dream is to live in a neighborhood where you know and talk to your neighbors. In the real world of cities and suburbs, this is often something that isn't as easy as it used to be. With a webcam, however, it's actually quite easy—in the global neighborhood of the Internet. For both personal and business relationships, human beings still feel better when they can see the face of the person they're talking to, and webcams can allow you to do just that.

There are certainly anecdotal stories of people who have taken advantage of this technology to hurt others by running scams, stalking others, or preying on children. In many respects, the Internet as a whole still has a bit of that frontier-town feel about it, and there are dangers.

Often when dealing with people you know over the Internet, security is, at best, a back-of-the-mind kind of thing.

Keeping it real (and safe)

While I have devoted entire chapters and parts of chapters to security, most of these deal with the technological side of security. But as long as I have you here (we all read the first chapter, don't we?), allow me to talk about security as it applies to you and your family. Often when dealing with people you know over the Internet, security is, at best, a back-of-the-mind kind of thing. Part of the whole fun of webcams, however, is that you can see and talk to people from almost anywhere. So what to do? Table 1.1 on the following page is a list of personal security issues with which you should be familiar.

Table 1.1. You should familiarize yourself with this list of safety guidelines.

Personal Security Guidelines
• Don't ever give passwords for any accounts to anyone over the Internet. The companies that you have accounts with either already know your password or they don't need it to administer your accounts.
• Children (especially those under 14 years old) should not be allowed to use a webcam to chat with others unsupervised. While many chat programs have filters for adult content, those filters will not stop something like a stranger asking your children where they live or where they could meet them.
• If you have a webcam broadcasting live video or snapshots from your house, be aware that if someone can find out where you live, that person could know when you are not home. This opens up the possibility of all sorts of nasty situations. My best advice is to not post a schedule of your comings and goings, don't have a time-stamp on your image that could provide cues as to the time, and show old images when you aren't on-line.
• Use common sense. While it may sound clichéd, many people on the Web are not hurt by high-tech villainy, but instead by their own carelessness.

Just what am I supposed to do with this thing?

Now that you're feeling all safe and comfortable, you may be wondering what you can actually do with a webcam. Well, that's what most of the last quarter of this book is about. But to summarize the kinds of applications, here are a few examples.

Imagine that you have a grandmother in Utah, while you live in Virginia. You write letters, you talk on the phone, and once every few years you might be able to afford a visit. Now suppose that both you and your grandmother have webcams (and a computer to hook it up to, of course!) and an Internet-access account. Suddenly, you can record video clips to send to one another, you can exchange video e-mail, or you might even use the webcam to video chat. In short, you could talk face-to-face every day for a month and only have to pay your ISP access fee (which for most people is around $19.95 a month) not a bad deal when compared to the expense of a month of long distance calls or a plane ticket to Utah.

Businesses can also benefit from webcams. One of the first webcams was of the lunchroom at a technology company. The company basically set it up to see if it was possible and thought it was sort of fun. The address for the webcam page was passed around and made it outside the company. While this company probably didn't need any help getting people to visit its web site, the added draw of this quirky little webcam certainly added to the overall traffic to the site. So while it was not originally intended to be anything but a toy for techies in the office, it turned into a cheap and original marketing tool.

Businesses can also use webcams to advertise services, showcase products, or simply improve employee morale.

Businesses can also use webcams to advertise services, showcase products, or simply improve employee morale by providing a view of the outside world from within their cubicles. Considering that most webcams cost under $50 and don't require much to maintain (within certain limits), a new webcam application can often be worth it, even if the official return on investment is negligible.

Webcams can also be used by students, allowing them to talk to classmates and even teachers about homework assignments. I've heard of some parents who home-school their children, creating virtual classrooms to allow their children more interaction with other students. Many day-care centers are even setting up webcams to allow parents to check in on their kids throughout the workday. I even know a busy marketing representative who set up a webcam at his office so that his children could check in on him throughout the day.

These are obviously just a few examples of how this technology can be used. I haven't even yet mentioned the plethora of web sites that revolve around people and their webcams.

Often used as a method of self-expression, these web sites often serve no definable purpose, but should instead be thought of as the cyber-equivalent of performance art. Thousands and perhaps even tens of thousands (no one's really sure) of people broadcast their lives over the Internet for everyone to see. While there is much buzz about the titillation factor (some sites broadcast *everything* in a person's life), many people do it not so much to exhibit their bodies, but rather the whole gestalt of their lives.

Many of you might think that this is in response to such films as *The Truman Show* and *EdTV*, but the phenomenon actually predates those movies by a good many years. I'll discuss some of the webcam pioneers and trailblazers in the next chapter.

Just like the physical world, the webcam community represents the whole spectrum of humanity, as well as encompassing a spectrum of applications and uses.

So just like the physical world, the webcam community represents the whole spectrum of humanity, as well as encompassing a spectrum of applications and uses.

After reading this book, I hope you'll experiment with a little of everything—that's the fun of it, right? While tools are usually only bought to accomplish a single task, the really cool ones are those you find yourself turning to for a variety of tasks. I believe webcams are that kind of tool, and after trying some of the things described in the following chapter, I hope you'll agree.

So where do I start?

Even if you are familiar with your computer's hardware, Chapter 4 is helpful in outlining how it all comes together from the perspective of using your webcam.

I'll spend the next two chapters giving you some background on the history of webcams and introducing you to the terminology used throughout the rest of the book. Many books stick a glossary at the end. But unless you want to spend all your time flipping back and forth, this doesn't make much sense. Chapter 3 contains all the terms we use throughout the book and some helpful definitions. Due to the numerous areas of technology I'll be covering, I really recommend spending some time reading Chapter 3; otherwise, if you're reading the book all the way through front to back, you'll soon be lost. Don't feel bad though. Some of the terms were new to me as well!

Chapters 4 through 7 cover the basics of what you'll need to get started using a webcam. Chapter 4 details what you'll need from your existing computer or what to look for if you are considering buying a new one. Even if you are familiar with your computer's hardware, this chapter is helpful in outlining how it all comes together from the perspective of using your webcam.

Chapter 5 focuses on the options surrounding the actual webcam. I'll discuss hooking up your cam, what to expect from the image quality, and how to judge the best value. You might think that, since all webcams use basically the same technology, they're all somewhat similar in what they provide. I'll explain why this isn't necessarily true and tell you how to judge whether the cam you're looking at is a deal or a lemon.

But what good would the best computer and webcam be without some software to run with them? Chapter 6 looks at the general classes of software that interact with webcams, what they do, what to look for, and what to expect. I can't explain the details of every package out there, but I'll give you enough of the broad strokes that you'll know what to look for when you need it and know how to judge what will best fit your needs.

Of course, webcams and the communities of webcam users wouldn't be anything without the connections between them. In Chapter 7, I'll explain the connections behind all the pretty pictures. Still trying to figure out 56K, ISDN, ADSL, and the rest of the alphabet soup behind Internet access? Well, I'll not only explain what they all mean, but I'll guide you through selecting the best choice for you and your needs.

In Chapter 11, I'll cover the webcam application that people seem to find the most fascinating: setting up your own webcam web site.

After that, I'll roll up my sleeves and get to work explaining to you in more detail how to get the most out of your webcam for different applications. Chapter 8 will give you the lowdown on capturing video with your webcam, including tips on how to set up your shot, how to adjust the lighting, and how to deal with problems such as movement and file compression.

In Chapter 9, I'll explain how to take your newfound skills and use them to record and send video e-mail. I'll talk about getting around the constraints of bandwidth, what format to save your video in, and how to make sure recipients of your video e-mail see your message the way you want them to.

After that, I'll get into the world of instant messaging and video chat in Chapter 10. This chapter will guide you through the maze of finding others to talk to, rules of etiquette, and the best way to set up your computer and webcam to chat with others easily.

Then, of course, to top it all off, in Chapter 11 I'll cover the webcam application that people seem to find the most fascinating: setting up your own webcam web site. I'll explain how to define a goal for your site, provide tips on how best to design your site around constantly updating content, where to find information, and how to get your site noticed.

Chapters 12 and 13 will deal with more advanced issues—not necessarily focused entirely on webcams—that are also part of living the "camlife." In Chapter 12, I'll talk about the technical aspects of setting up a business web site, including advice on how to sell the idea to your boss. Chapter 13 will give you some information about other kinds of webcams, everything from wireless mobiles to server-less stand-alones. I'll explain why and when you might use them and talk about other options for hooking up live video.

I'll conclude the book with two chapters on issues that I think you may need to consider. Chapter 14 will deal in more general terms with using webcams in a business environment, telling you how best to leverage it to your advantage. Finally, Chapter 15 is all about security: technical, physical, and personal.

As mentioned, I designed these chapters to be read either consecutively or independently; you can read the ones you want and simply use the others as a reference guide. For further reference, I have included an Appendix that lists on-line resources, including news groups, hardware and software manufacturer web sites, and webcam community web sites.

I designed these chapters to be read either consecutively or independently.

The Click Heard 'Round the World

Chapter 2

Jennifer Ringley's experiment

On a night back in the mid-'90s, with a click of the mouse, a person was able to view a web page that contained a mediocre black-and-white image of a modest college dorm room at Dickinson College in Pennsylvania. The image was updated every three minutes. Pretty standard fare for the day, as many people had already put up aquarium cams, window cams, and other sites devoted to displaying images of interesting (and not so interesting) public places. What made this web site special was that, as far as anyone knows, it was the first to focus on a person. Jennifer Ringley started the cam as a kind of experiment and kept it going for friends. She appeared 24 hours a day (though when the lights were

What made this web site special was that, as far as anyone knows, it was the first to focus on a person.

out, all you got was an image of blackness) and let the world see her at her best and worst.

Being first isn't all rainbows and sunshine though. The stories are varied, and Jenni has been quiet about it, but either the college pushed her to shut down the site (which was using the school's Ethernet connection) or she simply had to move and didn't have the proper connection at her new place. Either way, the black-and-white cam was in mothballs. A great idea can't be put down, however, and Jenni soon relaunched her site with a new color cam. She still intended the site to be a low-profile, "private" thing for friends. The other thing about a great idea is that it can't be kept private. Soon people from all over the world were logging on to the site, overloading her meager resources.

Many (and by "many" I mean men) were attracted to the idea of a woman living her private moments in the view of others, while some women felt empowered by the lack of self-consciousness displayed by this young college student. As her popularity soared, more and more problems cropped up. Several times the server hosting her site crashed from the sheer volume of hits. Hacks of her site and death threats soon followed. Jenni had tried to keep the site private, but growing interest from the public, a server crash caused by the sight of Jenni and a "friend" engaged in a liaison of a very private nature, and press coverage from as far away as Australia brought matters to a cusp.

As her popularity soared, more and more problems cropped up. Several times the server hosting her site crashed from the sheer volume of hits. Hacks of her site and death threats soon followed.

Jenni brought her site down again, though she relaunched it again in 1997 as a paid subscription web site. Subscriptions were $15 a year. So she made it rich and lived happily ever after, right? Well, not exactly. When she last spoke publicly on her setup to do all of this, she said she had two T1-lines and one T3-line. For those of you who don't know what that means, I'll put it in perspective: may small businesses get along just fine with *one* T1-line. As you can imagine, that costs quite a bit of money, as does the servers, cameras (she now has many throughout her living space), personal computers, and everything else that goes into making her site. Also, hackers try to get into her site hundreds of times a day, forcing her to actually have two servers and beefed-up security.

The future is not all bleak by any means though. She has become a bona fide celebrity, appearing on talk shows, TV dramas, and movies. The movie rights to her life story have been optioned, and there was talk of a pilot for a TV show based on her. She is instantly recognizable to millions, and her fans have a reverence for her that would be the envy of many a starlet. In addition, being on the Internet has insulated her somewhat from the bumps and bruises accorded to other mainstream-media celebrities by the tabloids.

Now, just a few short years later, there are thousands of people who also broadcast their lives for others to see. Why they do it puzzles many, and the reasons offered by those who do differ greatly.

Now, just a few short years later, there are thousands of people who also broadcast their lives for others to see. Why they do it puzzles many, and the reasons offered by those who do differ greatly. Jenni, in press interviews, has said that she does it to show the parts of life not normally seen in mainstream media, to strip away the pretense of celebrity, and to show that no matter who you are, life goes on for all of us in much the same way. She seems to enjoy the effect her site has on people, and says she plans to continue doing it until she's old and gray. She also is often publicly frustrated with those who don't understand what she's doing. In 1998, the head of a major cable company accused her of being "immoral." That same year, however, computer-publishing powerhouse Ziff-Davis voted her top newsmaker of the year in an on-line poll. Obviously, Jenni is a lightning rod for public opinion.

And none of this would have happened if not for the invention of the webcam. While you probably will not be planning on creating a site like Jenni's, certainly her tale amply demonstrates the cultural changes that can be caused by new technology. From a small circle of friends to worldwide recognition—not to mention her site being featured in a 1999 exhibit on fame and photography at the Museum of Modern Art in New York—Jenni's little experiment has had broad repercussions.

If JenniCam is the equivalent of the art of the realists, AnaCam is more like an impressionistic painting.

Pandora's box

Following closely on the heels of Jennifer Ringley's site, others also recognized the possibilities of webcams and began exploring them. Since they are relatively easy to set up on a small scale, many have found webcams to be an avenue in which to explore their own artistic visions. A perfect example of this is Ana Voog, a musician and performance artist. Her site is an extension of her artistic efforts. While Jenni pretty much shows her life as is, Ana chooses to push the envelope and uses the technology to bring you her life as an artistic extrapolation.

If JenniCam is the equivalent of the art of the realists, AnaCam is more like an impressionistic painting. While you may see her in a simple, unaltered video capture, more often than not her image has been processed through digital filters in the effort to achieve a certain effect, in a way letting you see her life as she sees it herself. Her main site is free, but she also has a paid subscription site for which she does on-line "shows" or performance-art pieces. Like Jenni and most other webcam sites, Ana also has on-line journals that provide information on her life and activities, giving the viewer a context in which to look at her images.

The exploration of artistic and creative impulses seems to be a common theme among people who have their own webcams.

One site I know of shows stark, mostly black and white artistic nudes of the cam operator, simply known as Kimberly. Her current career really has nothing to do with the art world, but she uses her cam to express her artistic skills and has built up a portfolio of work that would do many professional photographers credit. Much like the way in which the Internet has allowed people to publish their written work to millions for free, webcams have created an environment in which people feel free to express themselves in a visual medium.

After the inventions of the telephone and television, it was soon imagined that the two might one day be combined into a sort of videophone. In fact, videophones have been around—at least in prototypes—for decades.

Digital carrier pigeons

Of course, the history of webcams does not center completely on webcam Internet sites. From the beginning, the bulk of webcam users have used them to communicate directly with other webcam users.

After the invention of the telephone and television, it was soon imagined that the two might one day be combined into a sort of videophone. In fact, videophones have been around—at least in prototypes—for decades. But somehow the idea never really caught on. It took a new technology and the growth of a new subculture for the idea of a device that would allow you to talk to someone you could also see to catch on. The new technology, of course, was the modern computer, the Internet, and webcams. The new subculture was the group of people who had become accustomed to communicating with people via e-mail, on-line message boards, and instant text chat.

Once webcams became available, they started filtering through this subculture, which recognized them as a possible improvement on existing methods of on-line communication. People began using them to talk to other webcam users, either carry-

There really is no functional difference between the futuristic videophone imagined by your grandparents and using a webcam to engage in a video chat. The differences are mainly technological in nature.

ing on a conversation through typed comments or actually speaking and sending a live audio stream as well as the video. The constraints of existing technology soon became quite evident. The image was, at best, jerky, due to slow transmission speeds and technologically immature compression standards. The audio stream was even worse. It was obvious that, while a frame of video could be dropped without too much consequence, a second or two of dropped audio led to an incomprehensible stream of gibberish.

Despite this, webcam users carried on. Image and audio compression got better, computers got faster, and best of all, potential access speeds to the Internet were increased by faster modems and cheaper high-speed bandwidth solutions such as ISDN and cable modems. Standards for the transmission of audio and video between users were created. Today, we are on the brink of an explosion in digital communication.

The videophone, imagined by the pundits of the 1950s, is fast becoming a reality—though not as originally envisioned. While there are currently videophones that you can use in place of your normal phone, they have many problems, chief among them the ultimate technical boogieman—bandwidth. Your modern phone, which these videophones are based on, is still at heart an analog device. Your voice takes up most of the available bandwidth, and with videophones a still image is piggybacked on the signal.

But with a webcam, a computer, and at least a 56K modem, you can set up the equivalent of a videophone much more easily. There really is no functional difference between the futuristic videophone imagined by your grandparents and using a webcam to engage in a video chat. The differences are mainly technological in nature. The audio and video compression is much easier on a computer, as the processor is much more powerful than a chip that can be placed in a phone. Also, the stream of video and audio can be much more carefully managed over an Internet connection, making it better able to handle network traffic congestion.

The main difference, however, revolves around the growing subculture that has come to rely on their computers for communication. They're comfortable with the idea, whereas many traditional videophones failed due to a lack of public acceptance. It seems that, while the concept of talking to a person you can see intrigues us all, many people don't feel comfortable with it, and for videophones to work there would need to be wide acceptance of the technology. That's the great thing about chatting via your webcam: it is a matter of personal choice, and the infrastructure to support it is already in place. There are entire on-line communities based on matching up people who want to use their webcams to talk to each other. And as bandwidth speeds increase, more and more people will flock to video chat as an alternative to telephones.

Guessing where the future will take this trend is a losing proposition, but surmising that the computer will replace or incorporate traditional telephone traffic is not exactly risky. The question that companies are dying to figure out is when will it happen and how will it take place. And though engaging in a little crystal-ball gazing is interesting, the important thing is that, in terms of webcams and talking with others, the future is now.

As bandwidth speeds increase, more and more people will flock to video chat as an alternative to telephones.

Where do we go from here?

New uses for webcams are constantly being developed. Today, there are packages that use your webcam to incorporate pictures of your children in educational and entertainment software. When you consider the broader field of all digital imaging devices, you see that the applications are growing at an even faster rate.

Recently, a new type of convertible webcam was released on the market. The one made by Creative Labs can either be used as a webcam or a low-end digital camera and costs less than $150. This is the real trend to watch with webcams. Currently, the only real division between webcams and traditional digital cameras is price. Most quality digital cameras cost anywhere from $400 to $1,000, and the professional level models can run more than $5,000. That's a bit much for most people, and when you consider the types of images you see on your screen, the high resolution offered by normal digital cameras is overkill. In a sense, this explains the current market in webcam hardware. People are looking for lower-resolution, inexpensive cameras that are easy to set up and use.

Currently, the only real division between webcams and traditional digital cameras is price.

So for now, webcams are an ideal solution. But as digital cameras get cheaper—and they certainly will—look for more and more of them to be designed to fill the shoes of both webcams and digital cameras. As it stands now, most digital cameras can be used as webcams, but their physical designs are not usually amenable to being set on top of a monitor. Also, there's the issue of constantly connecting and disconnecting the camera from the computer. Look for camera manufacturers to solve both these problems by creating docking stations, much like those currently available for mobile computers. In fact, the aforementioned dual camera from Creative Labs uses just that approach.

What else can you expect? Once again, that depends on bandwidth. Depending on which of the high bandwidth options becomes the de facto standard and when it gains broad acceptance, webcams and the applications that interface with them will be called on to deliver sharper images with little or none of the jerkiness seen in Internet-transmitted images today.

The important thing to remember about webcams is that it's not a particular technology that makes them exciting; it's the concepts of communication and how they will change that will drive the market for the foreseeable future. No matter what happens with your camera, your computer, or your modem, people will always want to post images of themselves on the Internet, they will want to talk live with friends and family, and they will want to create videos to capture special moments in their lives. "Webcam" is the word for the device that currently does this, and that may change over time, but the ideas will remain the same.

The webcam pioneers I've talked about and others in the webcam community all seem to agree on one thing: webcams, in any of their myriad uses, are not about lenses, chips, or software; they are about communication.

The webcam pioneers I've talked about and others in the webcam community all seem to agree on one thing: webcams, in any of their myriad uses, are not about lenses, chips, or software; they are about communication. Whether it is a web site, a chat window, or a video file, these are all ways for us to breach the isolation that can come with a technological society. Think of the cultural changes that print, movies, radio, and TV caused. Then remember that the means of producing those media were traditionally in the hands of only a few. Imagine what a million Guttenbergs, Marconis, Edisons, and Bairds can come up with. Or, perhaps even better, what can a million Shakespeares, Orson Wells, or Lucille Balls come up with?

Interesting questions, to be sure. But for the rest of this book, I'll simply occupy myself with answering your questions on how to set up and use your webcam. Whether or not you become a famous cultural icon is entirely up to you.

Terminology: Talking the Talk

Chapter 3

The point of this chapter is to give you a frame of reference with which to approach the rest of the material in this book.

The terminology surrounding webcams specifically and computers in general can sometimes seem like an entirely different language. And like any discipline, there is a vocabulary unique to computers. I could fill up this entire book with words and definitions you could use, but that isn't what I'm after. The point of this chapter is to give you a frame of reference with which to approach the rest of the material in this book. These terms are generally united by some connection to the use of webcams. More general computer terms and their definitions can be found in the Glossary at the end of the book.

While you certainly can come back and look through this chapter at your convenience, I suggest you at least give it a brief run-through before going on to the others chapters.

CCDs are one of the two types of chips in digital video cameras. They generally provide the best image quality, though they are more expensive.

Application-sharing: Using a videoconferencing application to allow access to an application on a single desktop to multiple users.

Bandwidth: Term used to describe the speed at which data travels over a connection at a given time. Usually given as a theoretical potential maximum, actual performance is usually lower.

Cable modem: A special type of modem used to connect computers to cable TV systems to allow high-speed access to the Internet.

Capture: Refers to saving the information currently displayed on a screen or within a window. It can also mean the saving of image data from outside sources, such as a TV tuner-adapter card saving images from a VCR or webcam software "capturing" an image from a digital video camera.

CCD: Charge Coupled Device. Usually used in arrays, CCDs are one of the two types of chips in digital video cameras. They generally provide the best image quality, though they are more expensive.

Color depth: This is the number of distinct colors that can be detected by an imaging device and depends on the number of bits used to represent each pixel. A 24-bit camera, for example, can detect 16.7 million individual colors.

CMOS: Stands for complementary metal oxide semiconductor. It is the second possible type of chip used in webcams. CMOS-based cameras are cheaper and use less power (important for mobile users); however, they do not perform as well in low-light situations.

Dedicated connection: Provides Internet access 24 hours a day. More likely to have a static IP address assigned, it will probably use ISDN, cable modem, or DSL if a home user. It can raise the risk to your computer from hackers if no precautions (such as a personal firewall) are implemented.

Dial-up: An Internet connection that must manually be engaged, as in a typical 56Kbps modem.

Digital video camera: Also know as a webcam or desktop camera, these imaging devices are usually permanently attached to a computer, and those of the USB variety have no independent power source. They feature image resolutions usually no greater than 640 pixels wide by 480 pixels tall and are cheaper than digital still cameras, which are used to take much higher-resolution images and are usually used away from a computer.

Distance learning: The use of videoconferencing in education and training, usually combining audio and video feeds with other shareable applications.

Dithering: This can give images the illusion of either greater resolution or greater color depth. An example would be a TV screen where the image is made up of varying the intensity of three differently colored dots.

DSL: Digital Subscriber Line. A type of connection to the Internet over normal existing phone lines, but using different frequencies. Most consumers will use a form of DSL called ADSL, which stands for Asynchronous Digital Subscriber Line. This allows faster downloading than uploading, and as a result is cheaper than Synchronous DSL, which has the same data speeds both ways.

Filter: Term used to describe either a physical medium or software tool that alters an image.

Webcams feature image resolutions usually no greater than 640 pixels wide by 480 pixels tall and are cheaper than digital still cameras, which are used to take much higher-resolution images and are usually used away from a computer.

FPS: Frames per second. The measurement of performance with video applications. Full-motion video is 30 frames per second, meaning that 30 distinct images are displayed every second. A normal TV picture in the United States is made up of 60 interlaced images per second. The more FPS, the more fluid and natural looking the video.

Gallery: A collection of webcam images.

gif: Graphics Interchange Format. A standard format that computers use to store photographs and other graphics.

H.323: Industry standard for how videoconferencing data is transmitted across networks. It enables videoconference participants to take part in the same conference, even though they're using different applications.

Hyperlink: A piece of an electronic document, usually coded in HTML, that links to another document.

Instant messaging: Used to communicate (usually by text and occasionally voice) real-time with another individual. It also can be used to alert others to your presence on-line.

IT: Information Technology. This is the term used to describe computers, communications hardware, and any technology that deals with the science of storage or transmittal of information. The technical staff of most companies is called the IT staff.

The terms gif and jpeg are standard formats that computers use to store photographs and other graphics.

jpeg: Joint Photographic Expert Group. A standard format that computers use to store photographs and other graphics. Photos in jpg format can be compressed down to five percent of their original size, though some detail is lost in this process.

MPEG: Motion Picture Expert Group. This group determines the main standards for streaming video over a computer. (This group is described at length in Chapter 9.)

Pixel: Stands for Picture Element. A pixel is a single point in a graphic image. On computer monitors (and TVs), each pixel is actually composed of three dots—one red, one blue, and one green. With webcams, the number of pixels is determined by how many sensors are in the camera. Most webcams can produce an image with a total of 307,200 pixels. For comparison, high-end digital still cameras can produce images with millions of pixels.

> *Rather than the typical downloading of data, which requires a distinct beginning and end to the data, streaming video is the continuous transfer of information, usually video and/or audio.*

Resolution: The sharpness and clarity of an image. When speaking of screen resolution, this is determined by how many pixels are on the screen. When speaking of camera resolution, it is determined by how many pixels are in the output image. The resolution is usually measured by giving the number of pixels down and across, like rows and columns. So a typical 640 x 480 webcam image has 640 columns of pixels and 480 rows of pixels. Screen resolution is usually no higher than 72 dots (or pixels) per inch, while most printers print at 300dpi or above. This is why a large image on the screen can become quite small when printed.

Streaming: Rather than the typical downloading of data, which requires a distinct beginning and end to the data, streaming is the continuous transfer of information, usually video and/or audio. This is used in videoconferencing and streaming-video webcam sites.

tif, or tiff: Tagged Image File Format. A standard format that computers use to store photographs and other graphics. This is the most popular format for Macintosh computers.

USB: Universal Serial Bus. The USB ports on a computer are small plugs that USB external devices can connect with. It seems to be the standard that the computer industry is currently moving toward when connecting external devices.

QuickTime: A program made by Apple Computer to deliver streaming video. It is somewhat of a competing format with MPEG and works on both the PC and the Mac.

Videoconferencing: Sometimes called video chat. The process of having an on-line meeting between two or more computers at separate sites by using data networks to transmit audio and video data, and sometimes even application data. This requires that each participant's computer have a video camera, microphone, and speakers.

Video e-mail: E-mail that has a file attachment containing digital video. These attachments can be quite large, but can be used to convey information that plain text cannot. To play the video attachment, proprietary software often has to be downloaded or sent with the attachment.

Webcam: Any video camera whose output is available for viewing via the Internet or an intranet. This can be through a web site, video e-mail, or videoconferencing.

Web server: A device (usually an individual computer) that hosts web pages. Each server has a static IP address and usually a domain name. Web servers can either be a dedicated computer, or software running simultaneously with other applications on either a file server or desktop computer.

Video e-mail is *e-mail that has a file attachment containing digital video. These attachments can be quite large,*

Web site: A unified collection of web pages, usually organized into a hierarchy, and reached through a single domain name, such as www.website.com. Each page within the site has a unique URL, such as www.website.com/default.htm or www.website.com/search.htm.

PART 2:
The Right Tools For the Job

The Computer: Hitting the Performance 'Sweet Spot'

Chapter 4

Even with the best webcam in the world, you won't be able to do much if your computer hardware is too weak to drive all the utilities and tools for your applications.

Once you have decided that you want to set up a digital video camera, the first thing you need to do is know what you want to do with it. Do you want to videoconference over the Web, send video e-mail, or set up a webcam? This initial decision will help you then decide what you need in terms of computer hardware. If you're going to use the cam for video e-mail, videoconferencing, and other occasional uses, and if you have a fairly new (less than two years old, three at the most) computer, you probably won't need to buy anything else.

But if you decide that you want to set up a webcam, meaning a camera that updates images continuously on a web page, then you need to make sure you have the proper hardware to run your new endeavor. Even with the best webcam in the world, you won't be able to do much if your computer hardware is too weak to drive all the utilities and tools for your applications. For most webcams, you will actually want to have a separate computer to handle the webcam. This both frees up your machine from the task of feeding images from the cam to a web site and ensures that you can use your own machine when and how you want to.

This second computer can either be an older one (not more than three years though), or you can look for a lower-end new one or pick up a decent used or refurbished computer.

When buying a new one, figuring out what type of computer to purchase can be a difficult process. Even computer experts spend months wrestling with this question every time they decide to upgrade. In a way, the more you know, the more difficult the decisions are to make. One solution that many people resort to is overbuying. This normally entails purchasing the most expensive computer money can buy. The problem with this plan in relation to webcams is twofold.

For the most basic home-type of webcam system, you won't get any performance benefits from a high-end Pentium III or equivalent computer over a more basic Pentium II. The only difference is that you could be out a thousand dollars.

First, you would be surprised how little the system requirements are for a good webcam system, but I will go into that later. For the most basic home-type of webcam system, you won't get any added performance benefits from a high-end Pentium III or equivalent computer over a more basic Pentium II. The only difference is that you could be out a thousand dollars due to overbuying. And if you plan to dedicate a computer to the webcam, you might even be able to buy a cheaper Pentium II Celeron-based computer. Computers based on this chip cannot handle running multiple applications concurrently very well, but that should not be a major issue if you dedicate the machine solely to the webcam.

The second problem with overbuying is that normally computer companies don't offer a computer brand or model

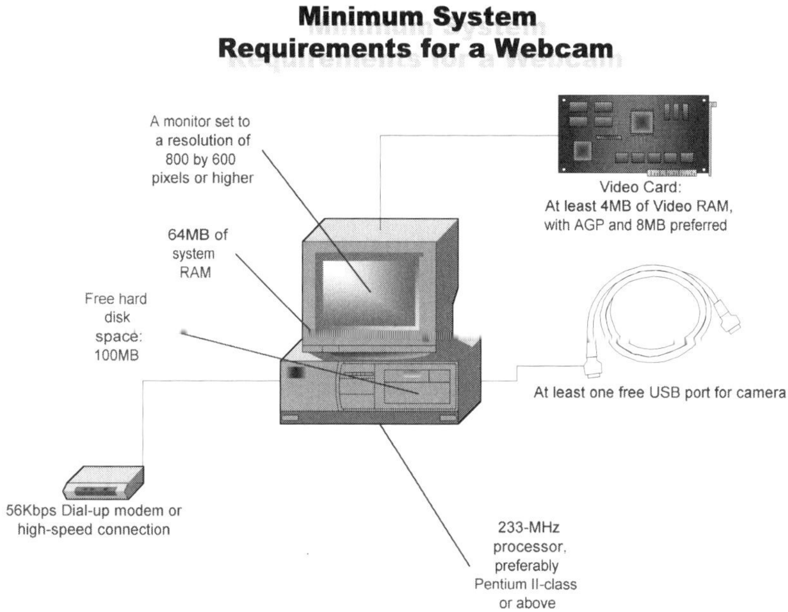

Figure 4:1. There are several factors to look for when purchasing a comptuer to support your webcam.

specifically designed for the type of setup required for a webcam. So users normally end up getting either a computer with too many or too few options. In the case of the latter, you'll have to upgrade the computer yourself, and in the case of the former, you're not only paying for accessories you may not use, but the additional internal components can actually interfere with the camera's operation.

The trick is to purchase the right computer for the task at hand. When you think about it, even that aging IBM PC long since exiled to your attic or garage is a pretty impressive machine, capable of doing millions of calculations per second. Now I'm not saying you could hook a webcam up to an IBM PC and expect it to work, but you won't need a monster PC either.

There are several factors to look for when purchasing a computer to support your webcam (see Figure 4.1 for the minimum system requirements). Again, while a dedicated computer is not 100 percent necessary, if you're running any type of business function through the system, it would be a good idea not to chance having another application such as an office suite crash, thereby killing the webcam in the process. If your goal simply is to use video e-mail or prerecorded or seldom-updated camera feeds, then you can probably have other applications running. But be careful sharing a system if the camera is anywhere close to mission-critical in its function.

Many people are surprised that the video card is actually the No. 1 factor that needs to be considered. But the truth is, since webcams are visual in nature, the graphics card is absolutely essential to cam operations.

Five factors

The five main factors when looking for a computer to drive your camera's functions are, in order of importance:

1) A good video card
2) Processor speed
3) Amount of system RAM
4) Serial or USB ports
5) Hard-disk space

Areas of secondary concern include the type of soundcard and the type of storage devices on the system other than the hard drive.

These recommendations are only for PCs. Macintosh computers, while suitable (and in many cases superior to PCs) for webcam work, don't have a lot of options. Since only Apple makes the computers, you take whatever you can get. But in the PC world, where there are hundreds of vendors, buyers can easily mix and match a system to suit their needs—whatever those needs may be. Although these recommendations are made for PC-users, Mac-users can still use them to narrow down the right system, just with fewer choices overall.

A good video card

Second in importance to the video card is the item that everyone thinks about first: the processor.

Many people are surprised that the video card is actually the No. 1 factor that needs to be considered. But the truth is, since webcams are visual in nature, the graphics card is absolutely essential to cam operations. A screaming, hot-off-the-line top PC with a low- to mid-range video card will offer substandard performance, while a mid-range system with a high-end card could be perfect when considering price versus performance calculations.

One of the tricks to determining which video card to purchase is to look at how much video memory the card ships with. At some point, the image from the camera is going to make use of video memory, and more is always better, especially when trying to capture motion. Some of the lower-end cards will have 2 MB of RAM, or perhaps 4 MB. The current standard is somewhere between 8 MB and 16 MB. Some of the best cards pack a full 64 MB of RAM, which is more than the total system RAM of mainstream computers just a few years ago. (Your old IBM PC would be so jealous.) Also, if you will be doing full-motion video, look for a video card that supports video compression through the card's own hardware. This will usually be mentioned somewhere on the box as natively supporting "MPEG" as well as "Indeo" and "DirectShow."

Another thing to consider is how the card attaches to the motherboard. Some cards are designed for a standard called PCI, while others are for AGP. The AGP cards are superior in performance to the PCI cards, because they offer better memory handling and a wider data channel into the motherboard. When trying to capture complex images with the webcam, the extra wide data pipe will come in handy. Some cards are designed for a standard called AGPX2, which is the second generation of AGP and is somewhat faster than

System RAM differs from video RAM in that the former isn't dedicated to video, and thus is less of a concern to webcam users.

the standard AGP cards. Also, there is the AGPX4 standard, which is even faster still, but the difference in terms of cam use is negligible. Most people just don't need such a large and fast video data channel, so there's not a lot of pressure to advance technology in this area at the moment.

Processor speed

Second in importance to the video card is the item that everyone thinks about first: the processor. Processors are getting faster, now pushing 1,000 MHz (or 1-GHz) and beyond. As a matter of perspective, that old IBM PC had a processor speed of about 2 MHz. But you don't need the fastest PC on the block or in the office to take advantage of webcam utilities. Most camera companies list the minimum requirements for their systems at around 133 MHz. But as I said before, you also have to look at the type of processor. There are different types to consider: Intel's Pentium II, Pentium II Celeron, and Pentium III chips, and AMD's Athlon being the main ones. The Pentium III and Athlon chips are optimized for streaming video, but other than that, in terms of supporting a digital video camera, there is little difference.

Despite being thrifty, you don't want to own a processor from the bottom of the technology barrel. New camera applications could be developed that require more processing power, and more power gives users and businesses room to grow if they suddenly find the webcam growing in importance and need to expand operations.

With no other applications running on the system other than the camera, a 166 MHz Pentium MMX processor would be the baseline in most cases. The preparing, capturing, and replication of images to the Web is a surprisingly simple process, and most cameras function at 100 percent on a 166 MHz system. Again, streaming video chews up more of your processor's resources, and you should plan accordingly.

If you are planning to use the webcam on a computer running other applications, particularly if those other applications are going to be operating at the same time as the camera, you are going to want more processing power. You're probably going to want at least a Pentium II or equivalent chip, because of the superior multitasking capabilities. On a heavily used system that also has to drive a webcam, consider getting at least a 300-MHz processor. One special note about the Pentium Celeron chips: because of a smaller amount of what's called Level 2 or L2 cache memory, the multitasking capabilities of computers with this processor are lower.

Amount of system RAM

Barring an inferno at one of the few RAM-making plants in the world (this actually happened a few years ago when RAM prices shot skyward), you should be able to add hundreds of megs of RAM to your system for around $100.

System RAM differs from video RAM in that the former isn't dedicated to video, and thus is less of a concern to webcam users. But make no mistake—RAM is highly important to having smooth overall operations on any PC. System RAM is a bit of a catchall for computers. One-minute system RAM can be devoted to running a word processor, and literally the next instant it can be assigned to help boost the performance of an Internet browser, assuming you are writing a document and surfing the Web at the same time.

Because the computer can read data in RAM almost instantly, most programs will try to load all or part of their code into RAM, which saves the system from having to access the slower hard disk or the even slower CD-ROM. In extreme cases, if too many programs are running, you will notice a slowdown as the system overspreads its RAM resources. In the epitome of bad situations under a Windows operating system, you might find the "blue screen of death" when overtaxing your system, meaning the screen switches to a bright blue color and you are presented with an error message. When this happens, you have no option but to reboot. Modern operating systems such as Microsoft's Windows 98 or

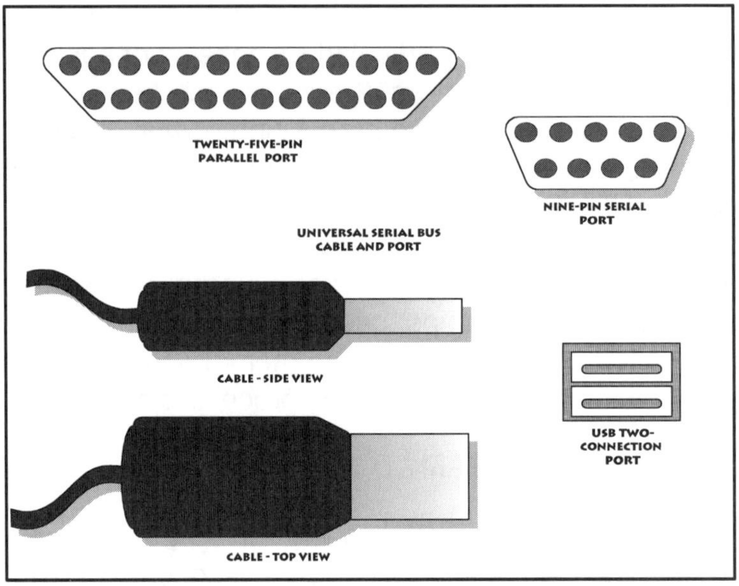

Figure 4-2. Drawings of various port types.

NT 4.0 are getting better at halting system crashes, but the best way to stop trouble is not to have it. And the best way to not have trouble is to have plenty of RAM for applications.

The good news is that RAM is cheap. The market in RAM fluctuates somewhat, but barring an inferno at one of the few RAM-making plants in the world (this actually happened a few years ago when RAM prices shot skyward), you should be able to add hundreds of megabytes of RAM to your system for around $100.

I've tested many webcams at various points during their history, and almost universally, they run with average performance on a computer with just 16 MB of system RAM. But they run flawlessly on a system with 32 MB of RAM. Again, this is based on what applications will be running alongside the camera. If you plan to do almost anything else with the system, consider getting at least 64 MB of RAM. This is actually the one area where you can afford to splurge when configuring a good system at a good price. Most systems ship standard with at least 128 MB of RAM, and some have 256 MB or more. If the camera is going to be the only thing your computer does, save some cash and drop the RAM down to 64 MB. It will guarantee good performance, and give you room to grow in the future.

Serial or USB ports

Most computers only have one or two serial ports, so if you're planning to use an older system with the webcam, first you're going to want to check to see if a port is free.

Another item you will have to consider when making a computer purchase is the type and number of ports a computer offers. Since a webcam has to exist outside of the computer—taking photos inside the computer gets boring pretty quickly—webcams are considered external devices. And all external devices have to connect into the computer using ports of some type.

The standard port type found on nearly every computer in existence is the serial port. The serial port is the smaller, square-looking plughole on the back of most computers. It can often be found beside the parallel port, which looks very similar but is much larger. You have to use a computer cable that screws into the back of the system to make use of the port. The cables are generally large and somewhat unwieldy.

Also, it is worth noting that not all serial ports are the same. They can be either male or female, which means the port has holes that point inward (the female port) or pins that stick out (the male port). While the serial port on the back of any computer always will be male, a peripheral device's serial port may be either male or female. If you get the wrong cable, you will end up trying to stick a male cable to a male port or a female cable into a female port. You need to be careful to have the right cable. Serial ports also are varied somewhat by the number of pins, though this is increasingly rare and is usually only seen on older computers.

Also, most computers only have one or two serial ports, so if you're planning to use an older system with the webcam, first you're going to want to check to see if a port is free. If not, you will have to see which external devices are using the ports and figure out if they can be disconnected or moved elsewhere.

As you might imagine, serial ports can be a pain. Thus, the computer industry invented a new way of connecting devices into computers called the Universal Serial Bus or USB connector. A USB connection is standard across all new PCs and Macs, and plugging one in is as easy as plugging in a light bulb. Many USB devices are even hot swappable, meaning you can plug them into an active computer, configure and begin to use them without having to reboot.

Another advantage to USB ports is that they're tiny. Three or four USB ports can squeeze into the same space as one serial port. And they are fast, too. Many USB devices perform better than similar models that use serial ports. So if your camera has USB connectors, do yourself a favor and make sure the system you want to use with it has a USB port. And since USB devices can be chained together, you will almost never squeeze a USB device out of your system when you add a new one.

There is, however, one pitfall to USB ports that many don't know about. Some older systems with the Windows 95 operating system shipped with USB ports. But the operating system and sometimes the system's BIOS do not support them.

There is, however, one pitfall to USB ports that many don't know about. Some older systems with the Windows 95 operating system shipped with USB ports. But the operating system and sometimes the system's BIOS do not support them. The easiest way to remedy this problem is to upgrade to Windows 98 and make sure you are running the most current version of the system's BIOS. I have no idea why computers shipped with unusable ports, but people considering using an older machine might want to check to make sure the USB ports on their system are more than mere decoration.

Hard-disk space

The final major consideration when looking for a computer to drive your camera's functions is hard-disk space. Most cameras are going to require about 75M of drive space for all the software applications, drivers, and various software components to run the camera. In today's world, where drive space is wonderfully cheap, 75M is a tiny drop in the bucket for most computers.

But you need to be aware that, depending on what applications you plan to run on the same system with the webcam and what type of webcam tools you will be using, you may need significantly more space.

For example, video e-mail takes up a lot of space—some are 10M or more each, depending on length and video quality. If you don't plan to ever delete your mail, like if you plan to archive it, those files can quickly eat up even a large drive in less than a year of heavy use.

Receiving a video e-mail without a soundcard is a fairly frustrating experience, because unless you can read lips, you won't be able to comprehend what anyone is saying.

If you plan to use the webcam to capture an image, then upload that image to a web page (you won't have to worry much about drive space). New images from the camera will replace older ones, and the bulk of the space that will be required will be on the Web server, not the local machine. For this type of operation, even a 2G drive should be sufficient.

If, however, you want to run the camera on a system with other applications, or use the local computer to archive video or even photographs, you might want to consider a 5G drive or greater. Many systems today ship with drives larger than 10G. With storage being so cheap, a super large drive is something you may want to consider, but there's really no need to fret over this option.

Second thoughts

Another secondary concern is the type of off-site storage device on the system. Yet here again, you can have a perfectly fine cam setup without one.

Of secondary concern is the soundcard. The reason this isn't a major concern is that cameras by nature do not require sound. You can have a perfectly successful business or personal webcam operation without ever giving a second thought to sound. Yet, the use of sound could enhance your overall use and enjoyment of the camera.

A few examples of where sound would be helpful are videoconferencing and video e-mail. Receiving a video e-mail without a soundcard is a fairly frustrating experience, because unless you can read lips—assuming the camera detail on the sending computer was set high enough to even make this possible—you won't be able to comprehend what anyone is saying.

A few cameras do ship with built-in microphones. Others ship with a computer microphone that plugs into a system soundcard. Microphone technology is pretty standard for computers, and one is nearly always as good as the next. I have seen some integrated microphones that ship with cameras that weren't very good, but mostly this was due to the poor design of the camera itself, not allowing for enough sound holes to reach the microphone.

A computer needs a soundcard to make use of the microphone. Since webcam applications are not exceptionally taxing on the sound side of processing, nearly any soundcard manufactured in the last few years will do. You won't be using a webcam application to listen for irregular heartbeats in patients or some type of mission-critical function; you will just want to have the ability to process and listen to sound. The main question is whether or not you want to include sound in your webcam operation, and whether or not you want to receive sound data from others. If the answer is yes, simply make sure there is a soundcard in the system you're buying. Even the most advanced soundcards on the market cost less than $150, so perhaps it would be best to err on the side of caution and get a soundcard, even if you don't think you'll use it initially. My guess is that you will find at least one application where sound will be helpful to you.

Another secondary concern is the type of removable media storage device on the system. Yet here again, you can have a perfectly fine cam setup without one. First, as far as drives go, most computers ship with either a CD-ROM or DVD drive. I included DVD drives here simply because they are becoming a standard option in almost all PCs. A DVD drive is absolutely unnecessary for a webcam setup. You can probably save some money configuring a system with a CD-ROM drive instead.

A CD-R drive can be used to burn your own CDs. This can be very helpful when archiving video e-mail and hundreds of megabytes of photos.

The only reason you need a CD-ROM drive is to install the camera and software onto your system. Since the CD-ROM can't write data, once you've handed off the information to the hard drive, the CD-ROM is not much help when considering storage needs. Therefore, you need to make sure you have a CD-ROM, but you can get the cheapest model available.

With CD-ROMs, cost usually follows the speed at which the CD-ROM can read data. The original CD-ROM could read 150-kilobytes per second, and now speeds are given as multiples of that speed. So a 2X CD-ROM reads at 300 kilobytes per seconds, and a 24X CD-ROM reads at 3,600 kilobytes per second. Even a slow 8X model (some are as fast as 64X or more) will only save you time when you initially install the software, and then only a minute or so. Basically, make sure your system has a CD-ROM and don't worry about much else.

For storage, some users might want to consider purchasing a CD-R drive instead of a CD-ROM drive. A CD-R drive can

When setting up a cam operation, the question always comes up whether a person should use a desktop computer or a notebook as the base for his or her video-capture program.

be used to burn your own CDs. This can be very helpful when archiving video e-mail and hundreds of megabytes of photos. Since a CD-R can hold 640M of data, you can save on hard-disk space by burning your own CDs filled with your cam data, and then removing it from the system. A CD-R drive also acts like a standard CD-ROM, so you don't need to have both on your system. One advantage to this method of storage is that, given proper care, a CD-R disk will basically last forever.

The problem is that the CD-R media is rather expensive. You can sometimes find them for about $1 a disk, but you will have to buy them in lots of a hundred or more to get that price. Also, CD-R drives can only write to a CD-R disk one time. After writing, the disk is "closed." You can't add any more data to it later, or remove data either.

Another idea you might want to consider for storage of your cam-based files is a floppy-drive replacement removable media device, such as the Zip drive from Iomega. Zip drives are pretty sturdy, and standard disks can hold up to 100M of data. The newest models can put up to 250M of data on a Zip disk. And unlike CD-Rs, Zip disks can be rewritten countless times. They're pretty fast, too. You won't get the same performance you would get out of a hard disk, but you will find that the results are pretty close. And Zip drives can be found for $100 or less in most stores. You can also get an external one that plugs into a USB port. Another option is a type of floppy-drive using technology from Sony called LS-120. The media for these drives can hold 120M of data, and unlike the Zip drive, floppy-drive replacements are backward compatible with standard floppy disks.

Of course, one of the major factors to consider in designing your overall webcam system is the way it will communicate with the outside world. This area is often a forgotten bottleneck of data. But it is so important that I have devoted all of Chapter 7 to connectivity issues.

In the corporate world, the difference between a videoconference station and a portable videoconference device could mean a lot less hassle.

Notebook or desktop?

When setting up a cam operation, the question always comes up whether a person should use a desktop computer or a notebook as the base for his or her video-capture program. Notebooks are a world unto themselves, and choosing to use one as your cam computer is rife with both helpful advantages and painful disadvantages.

The major issue in considering a notebook versus a desktop is your vision for the webcam. If you plan to use the system in one specific location, especially indoors with access to lots of power plugs, then a desktop is probably the way to go.

But this is an area that deserves some thought, since the advantages of having a portable base of operations are numerous. Consider the typical home user who only wants to send video e-mail to friends and relatives and has no plans to set up a live feed for Internet operations. With a desktop system, the user is more or less shackled to one room, probably a home office, den, or other type of family or computer room. That means that everyone who wants to send video e-mail has to go to the location of the computer, being that the cable on the camera is only, at most, six feet long or so.

If the camera were connected to a notebook computer, however, the only real restriction encountered when filming events for video e-mail are the battery life of the notebook. You no longer need to bring the new baby into the den so a distant grandmother can see the child. Also, the notebook can be brought outside just as easily as it can be taken to the

What I have seen in my testing with webcams on notebook computers is that often the images recorded by the camera are subject to motion blur.

dining-room table. Parts of outdoor sports events could be filmed, as well as picnics or a walking tour of the new rosebushes. Then, the laptop could be plugged into the phone line and video e-mail of the event could be transmitted.

In the corporate world, the difference between a videoconference station and a portable videoconference device could mean a lot less hassle. As long as conference areas are equipped with phone or other data-transmission ports, the notebook with the camera can easily be taken to meetings for the benefit of off-site employees. Of course, notebooks also take up less space than desktops, so in a space-constrained environment they are not only a good choice, but also at times the only one.

There are, however, two major disadvantages when launching your webcam operations on a laptop. First, notebooks are much more expensive than desktops. A notebook with near-identical specifications as a desktop computer can cost a thousand dollars more. And each feature you add—more system RAM, for instance—will end up costing more than if you were piling it into a desktop. The notebook market is not a poor man's game.

So if you absolutely cannot live without a portable webcam, then by all means you should look into getting a notebook computer, but expect to dig deep into your pockets. You might be able to save some money by purchasing an Intel Celeron chip instead of a Pentium as the notebook's processor. For the purposes of using a digital video camera, there is no significant performance difference.

The other disadvantage to using a notebook to drive your camera is subtler and involves the video card. Nearly 90 percent of all notebook computers use ATI Technologies video cards—actually, they're video chips inside of a notebook. Many older notebooks use the ATI Rage Pro chip (or one similar to it), the same chip that shipped with first-generation Apple iMac computers. Today, most are upgraded to the ATI Rage Mobility chip, which offers more video RAM and better performance.

ATI is a solid company, and their graphics cards are excellent as far as mid-range performance is concerned. They are not, however, among the high end. And since they basically control the notebook video card market, the best video card you can get on a notebook is going to give you less performance than the best card you can get on a desktop system.

This really will only come into play when you're dealing with full-motion video of high quality or complex images. What I have seen in my testing with webcams on notebook computers is that often the images recorded by the camera are subject to motion blur. When people wave at the camera, for instance, their hands oftentimes will turn into a blur, instead of being tracked all the way through the motion. The faster the people move, the more blur is generated. The video chip, and often the processor, simply can't catch up with the motion being recorded by the camera. Since most people are trained by movies and television as to how sharply focused camera images should appear, the motion lines are highly noticeable.

Those of you who remember the very early days of computing probably remember what shopping for computers was like in those days of yesteryear known as the early 1980s.

As with everything else regarding computer specifications, the decision to use a notebook or a desktop is centered on what applications are going to be used. Still images on a notebook come out just fine most of the time, and even full-motion video of something such as a board meeting, where most of the people are sitting still, turn out fine, too. It's just harder to configure a notebook to handle full-motion video capture and other high-end applications, and price is often exponentially higher. Like many products, you're going to pay a premium for convenience.

There are a few secrets to finding the best deal. One of the biggest is buying direct from the companies that make the computers.

Hunting the best deal

Those of you who remember the very early days of computing probably remember what shopping for computers was like in those days of yesteryear known as the early 1980s. Computers were actually sold in specialty stores, and not just little stores either. A trip to the computer store was like a trip to buy a car, complete with a giant showroom filled with the latest models. Customers were encouraged to test the computers, while salesmen working on commission tried to up-sell you into a model that was "close to 100-percent IBM PC-compatible."

I actually had two sales people on the phone with one another at different stores competing for my business at one point. The winner was willing to drop the price of an original IBM PC by $100 more than the competitor. I still ended up paying several thousand dollars for the box, but finding the best deal was a satisfying experience.

Today, you would be lucky to find a dedicated store anywhere where you could walk in and come back out with a new PC. Gone are the glitzy showrooms and slick salespeople. Computers are more like appliances today, sitting in neat rows in places such as Staples or Best Buy. For the most part, consumers have to know exactly what they want before they go computer shopping, because very few places will do any amount of handholding for new computer buyers.

There are a few secrets to finding the best deal. One of the biggest is buying direct from the companies that make the computers. Purchasing a computer over the Internet, by visiting a company's web page, is easy. And the shipping costs are almost always far less than having to pay a middleman. But whether you prefer to buy your computer in a department store or over the Internet, you have to make sure you end up looking at the components inside a computer and not the box itself. A computer is not made up of the fancy case; it's made up of all the individual components bundled inside the steel and plastic box.

I have a friend who was setting up a webcam in his home so he could keep in touch with his family, who lived across the country. His idea was to use video e-mail to send messages to his mother and father once a week and to receive messages back from them on his computer. For some reason, even though he only had a limited amount of money to spend, he insisted on getting a computer with a "brand name," in this case a Texas company that is one of the top-selling name brands in the world.

When he asked me to look over the specifications on the system he was thinking about buying, I advised him that the video card was too anemic for video applications and the hard drive too small for storage of many e-mails. I then went on line and found several systems that would suit his webcam needs and also fit into his price range. The systems I found for him had the exact same or better components than the system he had originally chosen.

But whether you prefer to buy your computer in a department store or over the Internet, you have to make sure you end up looking at the components inside a computer and not the box itself.

When I presented him with this information, he was highly upset with me. "I don't want to buy a computer without a brand name," he exclaimed.

I proceeded to show him that the components inside the computers I had chosen were identical or better than the components in the one he wanted to buy, and that the systems I had picked out were between $200 and $600 cheaper than his choice. I explained to him that that the name on the box really does not matter, if there were top-quality parts inside.

All this was to no avail. He was convinced that he wanted to get a system with a brand name. I tried to tell him that buying

Remember, the most important thing is that your system works for your intended webcam setup. Saving money is actually a secondary concern.

computers was not like buying jeans; a potential date was not going to be impressed with the brand name on his computer. But he didn't listen.

A month later, he came back to me again. His brand-name computer was not powerful enough to handle his webcam applications. His system was slow to the point of being unusable, and when recording long e-mails for his family, it crashed. He admitted to me that he had more or less wasted $1,500 dollars getting a computer not suitable for his needs.

Still wanting to help him out, I found some new components that could be installed into his computer, and for about $400 more he was able to bring his system up to speed. But $400 is a harsh lesson for most folks.

The most important thing you can do when looking for the best deal on a webcam computer is to plan ahead. Make a list of all the components you would like to have in your computer, based on the criteria I described earlier. Even if you don't know much about video cards, for instance, you can put something like "must have at least 8 MB of video memory" on the list.

Take your list to the stores where you think you might buy your computer. As you walk the aisles of systems, compare the listed features to your personal list. If a computer has more than you need, try to find one that more exactly matches your needs, because you don't want to overbuy. Ignore the names on the boxes.

Once you have narrowed down your choices to a few different systems, write down the system name and model number and drive to at least one other store. At the new store, first zero in on computers of the same make and model as your choices from the last place and do a simple price comparison. Then look around to see if there are any different models that might suit your needs.

The more stores you do this in, the more likely you are going to run into an unannounced sale or simply a really good deal. But don't settle for an inexpensive system that won't fill your needs. Remember, the most important thing is that your system works for your intended webcam setup. Saving money is actually a secondary concern.

One you have found five or six models that you think you might like to purchase, it's time to hit the Web. Here is where you are going to find your biggest savings. Finding the Web address of most computer corporations is easy. If you want to purchase a Dell computer, simply go to www.dell.com. If the model you are considering buying is a Quantex, go to www.quantex.com. Rarely will you find a large computer corporation where the name is different from its web address.

There are two disadvantages to purchasing a system on-line: shipping costs and time.

Most computer sites will have a field called Search. Simply type in the model number of the computer you just looked at in the store, and prepare to be surprised. When the direct price comes up, don't be surprised if it is $500 less than the same model sitting in the store. Again, cutting out the middleman is a real advantage most of the time.

While you are on the company web page, look at computers that are similar to the one you were looking at in the store. Web stores are not bound by the restraints of physical space, so there are almost always more models to choose from on-line.

There are two disadvantages to purchasing a system on-line: shipping costs and time. Most companies have an easy-to-use shipping form where users simply type in their address and the system will calculate costs. The most I have ever paid for shipping was about $98 for two-day ground service from across the country, and this was a very heavy

On-line shopping is probably 100 times safer than giving your credit card away to a waiter in a restaurant.

system. Even with shipping added in, the cost difference is almost always in your favor.

The other disadvantage to on-line shopping—in case you can't tell, I'm a huge fan—is time. When you buy a computer in a store, you can normally walk right out with it. When shopping on-line, however, at the very least you are going to have to wait for the computer to be shipped to you.

Oftentimes though, you will also have to wait for the system to be built. That's right, remember that lack of space constraints? The reason for that is because all those computers aren't sitting around. A lot of companies will get your system into the assembly line queue only after you have ordered it. At peak times, such as the December holidays, you can expect to wait as much as a month to get your computer built.

But having your computer built on the fly has its advantages. For one thing, you can configure a system exactly how you want it to arrive. Like you might want a standard system, but with a larger monitor and a smaller hard drive. A custom system is just a few clicks away.

Another question I get all the time is whether on-line shopping is safe. The answer I always give back is that it is probably 100 times safer than giving your credit card away to a waiter in a restaurant. I mean, how much do you really know about your waiter, other than if he was quick with the drinks? He or she could easily copy down your credit-card number, or carelessly leave it in a place where some unscrupulous person could see it.

Comparatively, the Web is very safe as an electronic commerce device. Even with no additional security in place, there is so much traffic on the Web that a hacker would have a hard time finding packets that contain credit-card information.

But there is security in place. When you actually get to the point in the transaction where you are giving your credit-card number, a little lock icon will appear at the bottom of your computer's browser. Depending on the settings, you might be informed that you are entering a secure server. That means when you enter your credit information and push send, your computer first encrypts the data before shipping it to the other side. The secure server, which sets up the encryption, is the only device that can decode the information.

So now you have the security of so much traffic on the Internet, plus the fact that your personal data is scrambled. It's like handing your waiter a credit card that can be read only by the credit-card machine—without any numbers on it for the casual observer to read.

I have bought thousands of dollars in various merchandise over the Internet and never once had a problem. As long as you trust the site you are giving your card to, I doubt you will ever encounter stolen numbers.

With a great deal of thought, a little bit of planning, and some common sense, you can create a system tailored specifically for the needs of your ideal webcam operation. It doesn't have to cost a fortune either. If you follow the steps outlined here, I think you will be happy with your results. Just be sure to take your time and get the right system for the right price.

I have bought thousands of dollars in various merchandise over the Internet and never once had a problem. As long as you trust the site you are giving your card to, I doubt you will ever encounter stolen numbers.

The Camera: The Medium is the Message

Chapter 5

Truthfully, webcam manufacturers can make them almost any shape they want.

Types of cameras

You may see ads in the paper, browse web sites, or visit your local computer store and see most of what is traditionally thought of as webcams. These can look like small flattened circles, spheres, or in some cases even weapons from a popular '60s-era sci-fi show. Truthfully, they can make them almost any shape they want. Now that most webcams draw electricity through their Universal Serial Bus (USB) connections, they don't even need power supplies.

In a sense, webcams go back to the days of the early Brownie cameras, which were basically small cardboard boxes that you slipped film into and opened a crude shutter to take a picture. Traditionally, a webcam is simply a small plastic box that has a lens and a small electronic light sensing device. The only really complicated component of the piece is the electronic "eye," which is one of two types of sensors.

The first type of sensor is called a CCD, or Charge Coupled Device, and the second is called a CMOS, or Complementary Metal Oxide Semiconductor. Either term is a mouthful, but CCD is the most commonly used on quality cameras today, having started back in the early 1970s. CMOS image technology is actually older, but only recently has the quality improved to where it can even be compared to CCDs. So why two different technologies? Well, simply put, CCDs are expensive to make and require dedicated equipment to make them. CMOS sensors are much cheaper to make and can be made on the same widely used equipment that produces common computer chips, such as Intel's Pentium processors.

While CMOS cameras are certainly easy on your pocketbook and provide acceptable resolution—especially for a webcam—my advice is to steer clear of them.

There are even different types of CMOS sensors: active and passive pixel sensors. Active pixel CMOS sensors are better, but again are more expensive. So what should you look for? Well, there is a lot of talk about how CMOS sensors now provide the same image quality as CCD sensors and are cheaper to produce, which means cheaper digital cameras. Due to the nature of the chips, however, CMOS sensors will always behave poorly in low light. With every receptor on a CCD sensor, you get 100 percent of the light that hits that spot on the chip; with CMOS chips, you only get a fraction of the light that hits the receptor in your image. In addition, because of how the cameras are put together with the different types of sensors, CCD sensors perform better on video applications, and its images require less processing than CMOS images.

All in all, while CMOS cameras are certainly easy on your pocketbook and provide acceptable resolution—especially for a webcam—my advice is to steer clear of them. Most people don't use their computers surrounded with the equivalent of enough light to match that of a sunny day, and as a result, they will always be unsatisfied with the output from CMOS sensor-based cameras. On the other hand, almost all quality digital cameras, webcams, scanners, and even common video cameras use an array of CCDs to convert light into a digital image.

In the world of webcams, you will be dealing with devices that produce images of only 640 by 480 pixels at maximum, which is one of the reasons that webcams are notorious when it comes to low-light photography.

On a top-of-the-line digital camera, for example, you might have a CCD with 3,000 rows and 2,000 columns of photoreceptors. This gives the camera a resolution of 3,060 by 2,036 pixels, and the resultant images would contain more than 6 million pixels. If you were to print that image out at a respectable 600 dpi, you would have an image that measured 5.1 inches by 3.4 inches.

In the world of webcams, however, you will be dealing with devices that produce images of only 640 by 480 pixels at maximum. This means fewer receptors on the sensor and a darker picture, which is one of the reasons that webcams are notorious when it comes to low-light photography.

Some folks get around this by using a traditional video camera coupled with a TV tuner card for their PC. With this combination, you can get much better images in low light, due to the more sophisticated nature of the CCDs in camcorders. If you already have a camcorder and would like to use it instead of the type of digital video camera discussed throughout the book, feel free. Those TV tuner cards are only a little more expensive on average than a digital video camera. If you do choose to go this route, you can still use it with most of the applications talked about in this book. The software that comes with your TV tuner card should provide many of the same functions of webcam software, or you can buy applications that will meet your needs.

If you were the adventuresome type and had some time on your hands (as well as a degree in electrical engineering), you could even make your own webcam with parts available from catalogs and electronics stores. After all, a webcam is, in many ways, even simpler than traditional 35mm cameras. There's no film, no aperture settings, and no need to replace batteries—only a lens, a CCD, an analog-to-digital converter, and a way to take the signal to the PC. Of course, then you would also have to come up with a software driver and things like that. So for now, why don't you just take a look at some of the things to consider when buying a webcam?

A webcam is, in many ways, even simpler than traditional 35mm cameras. There's no film, no aperture settings, and no need to replace batteries.

With any computer peripheral, quality can vary extensively, and price is not always the best indicator. Digital video cameras are no exception. While the technology and electronics involved may even sometimes be the same from different manufacturers, the quality of your experience with the cam will still depend on a number of factors.

These factors fall into three categories: connectivity, image quality, and value. By connectivity I mean the type of physical connection between the cam and the PC, the software drivers that allow the cam to work with your operating system, and to some extent the ease-of-use of the graphics software that comes with the cam.

Image quality is a pretty simple term but can be influenced by a complex variety of things. The most important piece of advice I can offer is: Do not buy a cam until you've seen examples of the images it produces, if possible. Short of that, make sure that whomever you buy it from—whether a bricks-and-mortar retail store or an on-line retail site—offers a good old -fashioned return policy. People tend to buy imaging products, such as cams, scanners, and digital cameras, based on price and the promised image resolution. But color quality and performance under a variety of lighting conditions can be far more important, depending on what types of applications you use.

As for value, I'm not just talking about price. In the computer industry especially, heavily advertised retail sales are driven by what products will soon be replaced by a newer model. This can be either a curse or a blessing. Sometimes the older model will meet your needs, and sometimes it will be nothing but a lemon. Since cams have a minimal amount of parts that degrade or wear over time, you can plan on keeping whatever you buy for a while—that is, as long as the manufacturer continues to support the product with updated drivers, or even new ones in the event of the release of new operating systems.

Most cams come with a variety of simple graphics tools and utilities to get you started. If you only plan on using the cam occasionally, these applications will probably be your primary tools when using your cam. For businesses and home users who already have the graphic programs needed, however, these simple tools will most likely not prove very useful. You might even be able to catch a price break by looking for cams that don't come with a software bundle. After all, you should always try to pay only for what you really need.

Connectivity used to be a much more important factor when buying external peripherals.

Connectivity

Connectivity used to be a much more important factor when buying external peripherals. You had serial and parallel ports, SCSI devices, and those devices with proprietary interfaces. Now almost all external devices are connected to your computer via an interface called USB. The majority of computers sold over the past few years have two

USB ports, thus making it fairly simple for everyone. But there still are a lot of older computers out there. If you have an older computer, then the host of new USB cameras are not going to be appropriate—unless you have an add-in or option card with USB ports. If possible, USB is definitely the best choice, as it is the simplest to set up and has become the de facto standard.

To use USB-enabled cameras, you must not only have a computer with USB ports, but you also must be using an operating system with drivers for USB. Microsoft Windows NT 4.0 and older versions of Windows 95, for example, don't support USB. With operating systems from Microsoft, the best choice for USB support is Windows 98 and Windows 2000. Those who use the Mac platform can also use USB devices, as long as there are software drivers specifically for the Mac platform. For those with UNIX-based systems, and those running flavors of the Linux operating system, you have to deal with less driver support and connectivity issues with older systems.

Now, you may feel limited by the length of your USB cable. After all, there is a lot of world outside of a nine-foot radius from your computer. To answer these needs, many manufacturers have developed wireless and stand-alone cameras. More information about these cameras can be found in Chapter 14.

To use USB-enabled cameras, you must not only have a computer with USB ports, but you also must be using an operating system with drivers for USB.

Image quality

I already talked a little bit about how the camera achieves pixel resolution by the number of photoreceptors on the CCD or CMOS sensor, but that's just the beginning of looking at image quality. Most webcams can deliver a maximum resolution of 640 by 480 pixels, but that doesn't mean that images from the cameras are all the same—far from it, in fact! Images from webcams made by the same vendor can even have sharp differences in quality.

This is due in large part to differences in components and software. For example, some manufacturers may have a low-end and high-end model that look nearly identical, but one will have better quality components, such as a CCD instead of a CMOS sensor.

To get the best color accuracy possible, not only should your camera have software-based color-control tools, but you might want to take advantage of either the color profile capabilities of Windows 98 or third-party calibration tools.

The main differences between the images of different webcams is color quality and light sensitivity. Light sensitivity is the most important difference, since nothing can be done to fix it; it either is or isn't a good camera in low-light environments.

Color quality, on the other hand, has the potential to be fixed. A good webcam will have software that allows you to adjust the color values of your images and store those values, so you don't have to reset it every time you want to use it. This is a feature found in many name-brand webcam packages such as Kodak, Creative Labs, and Intel, and it's completely missing from those $24.99 specials you see from companies you've never heard of.

To get the best color accuracy possible, not only should your camera have software-based color-control tools, but you might want to take advantage of either the color profile capabilities of Windows 98 or third-party calibration tools. These can help make sure that the colors your camera sees, your monitor displays, and your printer prints are all in the same ballpark. This is not hugely important, but to judge how accurate your camera is, you should stick a mirror right next to your webcam and look at the difference between how you look in the image from the webcam and how you look in the mirror. You might be surprised how far off your color values actually are.

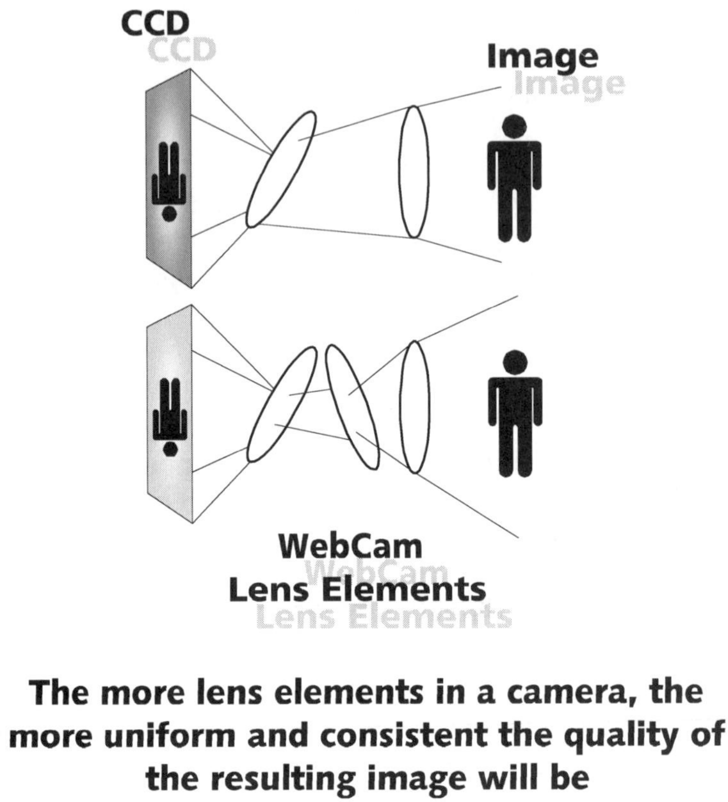

Figure 5:1. Webcam lens elements and sensor array.

So for performance in low-light environments, choose a CCD-based camera and look when you are shopping to see how many "lux" are required for good images from the camera. This should be listed somewhere on the box or on the manufacturer's Web page. Lux is a term used to describe light levels, and the lower the number your camera can work with, the better your images will be.

Other factors can also color your webcam experience. Low-end cameras often have only two or three lens elements, while higher-end models have as many as four. Both the number and the quality of lens elements are important, but this sort of information is rarely listed on the side of a box somewhere. Just know that if the image from the webcam is fuzzy or dark around the edges, the camera probably only has two lens elements. Fuzziness can also be caused by CMOS sensors, as they have more "noise" in the image than CCD-based cameras.

Value

Equally important to the camera, especially for the beginner, is what comes in the box along with the camera.

Obviously, you want to get the best camera for your money. The easiest way to do that is to look at the features I've mentioned so far and find the best deal you can on a camera that has the specifications you think you'll need. Equally important to the camera, especially for the beginner, is what comes in the box along with the camera. Are there photo-editing tools, and are they ones that will do what you need? What about an application to record video e-mails? Does it let you send the attachments with your own mail client, or does it do it separately?

As you can see, the applications that come with these packages can often be as difficult to evaluate as the cameras themselves. A good rule of thumb is to check to see if the applications are either full or limited versions of well-known programs. For example, a camera might come with a limited edition of the popular Adobe Photoshop or CuSeeMe Network's CU-SeeMe videoconferencing software. These are names well recognized in the industry and have whole companies devoted to making them work well by providing support. But I have also seen many vendors supply cameras with either proprietary software or software from small, independent companies. While these certainly may work wonderfully, chances are pretty good that you'll have problems. Small independents are great for making movies and coffee, but for something as complex and often difficult to use as software, my advice is to stick with the big brand names.

Also, many companies will include an offer for some kind of Internet service from a company with which they have a partnership. Beware of these offers. As laid out in the chapter on connectivity, the issue of who your Internet provider should be is too important to be decided by the last advertisement you saw. After you look at your options, if you decide to go with the company that has the included offer, feel free to take advantage of it. And if you decide against it, I can almost guarantee that the company you do decide on as an Internet service provider (ISP) will have a very similar offer for you.

So when you evaluate which camera to buy, my advice can be summed up as: look at the specifications, but look beyond them as well. The most important piece of wisdom I can offer you is to not buy a camera sight unseen—or if you do, make sure that you can return it fairly easily. The proof is in the pudding, as it were, and while digital images can be changed or modified any number of ways, quality can't be faked.

When thinking about installing a webcam, the old adage about measuring twice to ensure that you only have to cut once comes to mind.

Measure twice, cut once

When thinking about installing a webcam, the old adage about measuring twice to ensure that you only have to cut once comes to mind. After having done the research on webcams, after picking which one you want and bringing it home, now comes the fun part—hooking it up!

Before you do that, however, it's best to take a moment and think about your setup. Where do you want to position the camera? Is there proper lighting in the area where you will be using it? Have you read the installation instructions? Do you understand them? For answers to questions about where to put your camera, check out Chapter 8, which also describes how much and what kind of lighting is best, among other things.

The instructions should be fairly simple to install a webcam. As most of them connect to the PC through a USB cable, they are what's called "plug-and-play." This means that the computer will sense that a new device is connected and will ask you to provide the proper software drivers. The webcams themselves probably can be installed in under five minutes. The applications that go along with them, however, can take quite a bit longer. Your instinct will probably be to immediately start experimenting, but do take a moment and at least browse through your software's manuals. In the long run, the knowledge you pick up from that will enable you to enjoy using your camera and applications even more.

And on that note, flip a page or two to the chapter on applications. After all, a webcam is useless without the software that turns raw images into your own personal creations!

For answers to questions about where to put your camera, check out Chapter 8, which also describes how much and what kind of lighting is best, among other things.

The Software: Bringing It All Together

Chapter 6

There are three components vital to any webcam operation. The first two I have already talked about. To recap, the first thing you will need is, of course, the camera itself. Secondly, you need some type of computer hardware to host your setup. But the third component is as vital as the other two, though it is often overlooked: you need software to provide the brains for all that brawn.

No matter what you do with your webcam, you will need software to drive everything that you do. A webcam is not like a standard camera, or even a digital photographic camera. Disconnected from a computer, it's little more than a paperweight. Even with the best computer in the world

No matter what you do with your webcam, you will need software to drive everything that you do.

and the most advanced webcam, you won't be able to do anything without software.

No book could ever hope to explain the nuances of every piece of webcam software on the market. Whatever software you choose, you will have to roll up your sleeves and learn the ins and outs of your choice. Just remember that there are some really nice programs available that don't cost a thing. Many of these are available on-line. But no matter what software you choose, you will need a basic understanding of how webcam software works, and you will need to be able to determine if your chosen software package can fulfill the webcam vision you have in your head. That is what I will attempt to explain in this chapter.

The most basic of all webcam functions is imaging. Almost every webcam that you can buy has bundled software that allows users to create images.

Imaging

The most basic of all webcam functions is imaging. Almost every webcam that you can buy has bundled software that allows users to create images. In this mode, the webcam is most like a regular camera. This is basically the simplest mode of camera operation. Generally, a user simply stands in front of the camera and pushes a hot key on the keyboard or on the camera itself. Although you will not notice any movement, the camera will take a picture of whatever is in front of it. Normally, the image is then displayed on the monitor and a user has the option of keeping it or discarding it and trying again.

There are several advantages to taking a photo with a webcam. First, you don't have to worry about film. You basically have unlimited tries to get exactly the right shot. And, like an instant camera, you know at once whether the image

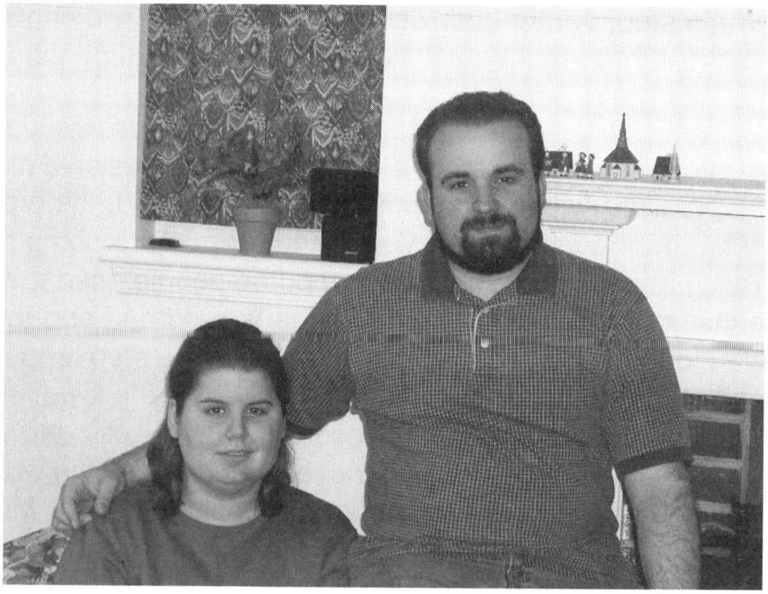

Figure 6-1. This image was captured with a webcam. Notice the quality is good, though not perfect.

is one you want to keep. In this respect, the webcam is literally instant, whereas an "instant" camera requires you to wait several minutes for the film to develop (and very expensive film at that, I might add).

If you have an inkjet or similar printer, you can load photo-quality paper into it and print your images. When holding an actual photograph beside one that was taken with a webcam and then printed on a nice printer, it's very difficult to tell the difference. A trained eye can tell by examining the lighter areas of the photograph, because on a real photo you can't see the dots that make up the picture. But modern printers do a good job of faking this, especially if you use the best photographic paper.

The true advantage to using a webcam is when a user wants to manipulate images on the computer, or otherwise use them with the computer, such as on a web page. Before webcams went mainstream, this meant taking a photo using an actual camera and then having the film processed. Instant cameras

With a webcam image, what you see is what you get. If the image on your screen looks good, it will look just fine on your web page. It's that easy.

saved a step, but the resolution often was not good enough to make it through the next stage, which was putting the images into the computer.

You see, even after you have a gorgeous, professional photograph taken with a 35mm camera in your hand, the problem is that the image is in your hand, not on the computer screen. To get images into the computer, people need to put the image into a scanner. The image is recorded and then saved to the computer. But this extra scanning step tends to hurt the image. Scanners, especially those on the lower end of the price spectrum, do not record colors exactly as they are in the photograph or as your eye sees them. In the worst cases, scanners can blur or distort complex images. Most often what happens is that darker areas on the photograph turn black, or lighter areas lose all sense of color balance.

With a webcam image, however, what you see is what you get. If the image on your screen looks good, it will look just fine on your web page. It's that easy. And a good webcam will come with drivers that allow you to configure things such as color balance, brightness, and contrast to ensure that your images are as true to life as possible.

Normally, when the camera software saves the photo, it will give it a graphical file extension. The most common are .gif, .jpg, and .tiff. So if you call your photo "grandma," it will get saved on your hard disk as grandma.jpg.

File formats can vary on PCs, but most imaging applications save images as .jpg format. This is because the .jpg format offers the most compression, which means less drive space is taken up by the photo. You could have thousands or even tens of thousands of photos in .jpg format in a single gigabyte of hard-drive space.

With small file formats and a real point-and-click ease of use, lots of webcams will be popping up in unlikely places. Even I was surprised last week to find a webcam in my gym. Now, the gym is hardly a place where intellectuals gather to spar with one another. And the staff, while extremely beautiful, are not exactly nuclear scientists. But webcams are so easy to use that one even has become an essential part of the operation of my iron-pumping organization.

When people register to use the gym, the attendant has them stand in front of the webcam and pushes a button to capture their image. Then, when members visit the facility in the future and scan their keycard, their picture appears on the computer monitor. At a glance, the attendant can tell if they are actually who they say they are, and there is no need to carry additional identification. The system is so simple that even the muscle-bound blond man who always appears lost in thought and the young girl who I've never seen not talking on the phone can use it. And isn't that the true test of any technology trying to break into the mainstream?

Streaming video

Making streaming video is almost as simple as taking a still image.

Although most people don't know it, streaming video is a close cousin to still images. You see, a webcam is more like a camcorder than a still-photo camera. It was designed to gaze out into your world with its unblinking eye. When you tell it you want to take a picture, it basically captures one frame of what it is watching and turns it into a picture.

Making streaming video is almost as simple as taking a still image. Since the webcam is basically designed to be able to capture streaming video, most cameras come with software that can do this.

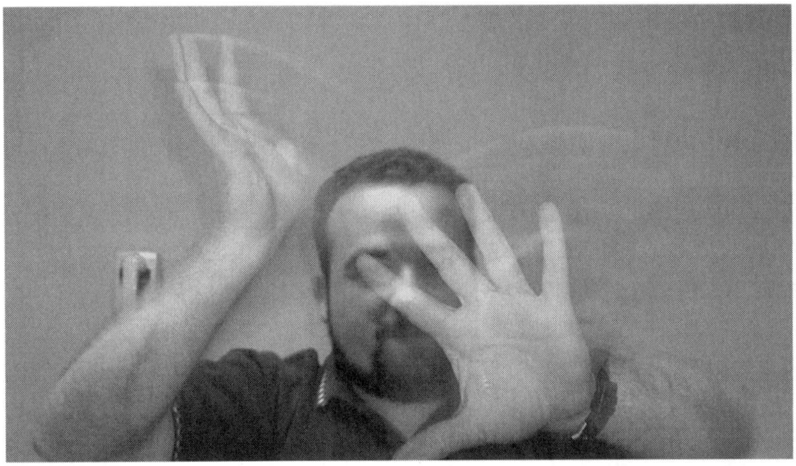

Figure 6-2. An example of motion blur when image quality is set at low.

The major difference between the capture of still images and streaming video is that the quality of streaming video is going to be more dependant on the hardware that is hosting the webcam operation and is going to take exponentially more space than a still image.

No matter what software you choose to create your streaming video, there's going to be a setting on it somewhere that lets you set how detailed you want your video to look. If you set the quality too low, even the slightest movement will result in motion blur. To some extent, setting the quality to lower levels could also lower the sound quality, if you are making a video with sound.

So why would anyone not tell the software to record images at the highest possible level? Isn't detail important? Sure it is. But the problem is that quality comes at a cost. And that cost is hard-disk space and bandwidth.

Take this example. A few years ago I was attending the PC Expo trade show in New York. As I passed one company's booth, I was enticed to come in with the promise of sending a streaming video of myself back home to loved ones. I was intrigued. Always anxious to play with a new toy, I accepted the offer and was led to a computer with a basic webcam mounted on top the monitor.

Once the technician got the system set up, he performed a little television countdown routine and then pointed to me as he activated the camera. I fumbled through a little presentation about how much of a good time I was having, gesturing to the massive show floor stretching out behind me. A quick, "I love you" followed and the video was finished.

Yet there was a problem. The technicians at the booth had used their software to set the image quality to the lowest level, figuring that if they did hundreds of these little presentations a day, they would need all the space they could get. But I had not stood still during my presentation. I had gestured to the show floor behind me and blown a small kiss. The camera could not capture my motion using the lowest quality setting, so I turned out looking like a huge blur.

The technicians quickly assured me that this problem could be fixed and used the software's quality slider bar to crank the image quality all the way up to the highest level. I then redid my presentation, which lasted about 30 seconds.

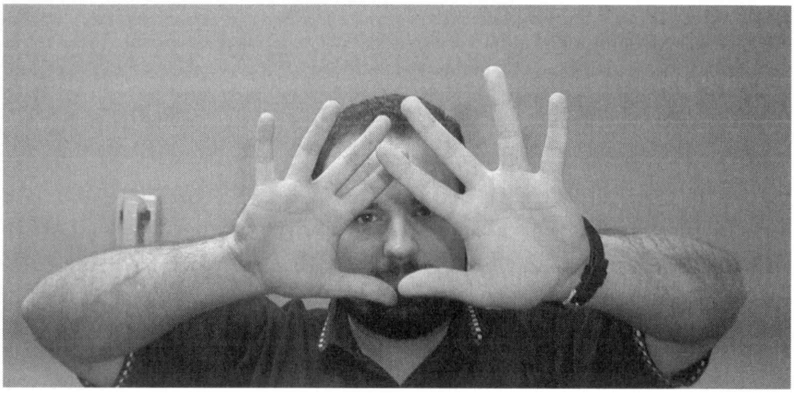

Figure 6-3. Roughly the same image as Figure 6-2, only without motion blur.

When I finished, I looked at the video and, sure enough, I was sharply in focus, even when gesturing, blowing kisses, and generally not sitting still. When they tried to e-mail the video, however, I noticed that it was taking a very long time for the message to get through. Curious, I asked them if I could examine the video file on their hard drive. Since at trade shows the man with the media badge is always right, they said that my probing was no problem.

Well, guess what? The high-quality, 30-second video I had created was 11.5 megabytes in size. In other words, it would have taken about eight 1.44MB floppy disks to hold it. I cringed when I thought about that huge message streaming through cyberspace toward my fiancée's computer. Using her 56K modem, I knew it would take a very long time to download the entire thing. Thankfully, the fact that I was thoughtful enough to send her a nice note overrode any anger she felt at wasting 30 minutes trying to download her mail.

So remember, when producing streaming video, be sure to experiment with the quality versus space slider bar on your software. If you're sitting still, a lower setting will probably be all right. Most people will opt for somewhere in the middle, with the bar perhaps slightly favoring the space-saving side. Just remember to sit still and keep the kiss blowing to a minimum.

Nothing has changed the world of computers more than e-mail.

Trying a webcam with e-mail

Nothing has changed the world of computers more than e-mail. Today, when you make business contacts, they're just as likely to give you their e-mail as they are their telephone number. And a webcam can be an important part of enhancing e-mail.

First off, you have to realize how important e-mail has become. It's practically the backbone of many businesses and an anchor in an increasing number of people's personal lives.

When people ask me how important e-mail is in my life, I have two stories to tell them. The first is when I wanted to do freelance writing for the newly launched on-line section of *The Washington Post.* Since it was a brand new section, I figured I would offer my services by sending the editor an e-mail. To my surprise, he e-mailed me back an assignment that I quickly completed. (Incidentally, that first article was a description of interactive webcams on the Internet.) A few days later it ran on line, and a few days after that a check appeared in the mail for me.

E-mail is practically the backbone of many businesses and an anchor in an increasing number of people's personal lives.

This continued for more than a year, with me getting assignments and sending them to the editor over e-mail. I had established a profitable business relationship and never once met the man who was signing the checks. When we finally did meet, we had a nice laugh about the whole thing. Today we're good friends, although we're both still slaves to our e-mail.

The second story happened when a friend of mine wanted to start up—of all things—a weekly trade publication. Since we wanted to cover the computer-game industry, we figured most of the people in that business had e-mail. So when we founded *Game Industry News*, we produced the entire thing on-line and published directly to our subscribers' e-mails. Today, four years later, we have a detailed support web page, but have never once produced a traditional issue on paper. And profits couldn't be better.

So now that I've established just how important e-mail has become, let's talk a bit about how a webcam can fit into the entire process.

Describing your new baby over e-mail is one thing, but sending a photograph of the new little person in his or her holiday outfit is quite another.

The ability to send photos through e-mail adds a visual element that is missing in regular text mail. No matter how great the writer, providing a description of an object or place that's better than a photograph is a difficult task. I mean, describing your new baby over e-mail is one thing, but sending a photograph of the new little person in his or her holiday outfit is quite another. In the business world, being able to see exactly what object the boss wants to order is helpful and eliminates mistakes.

Since the software driving the camera saves images in standard formats, sending them to others with e-mail is simple. You just press the attach button on your e-mail, find the file on your hard disk, and then select it. To make things simple for you, try creating a special folder where you store all the images generated by the camera. Your camera software can be configured to save all images into that special folder by default, and some software packages will even figure out that you want to do this once it detects that you have saved more than one image in that location.

Once the file has been selected, most e-mail programs will display a paperclip or some other symbol near the subject line that represents the fact that the image will go out with the e-mail. On the other end, the receiving party will either see the image imbedded below the e-mail's text or will see an attachment symbol and can click on it to view. Any computer with an Internet browser can view photos stored in standard formats, just like when looking at them on the Web.

The only thing you might want to take note of when sending photos this way is that it may take a bit longer for the e-mail to leave your computer, and your system might appear to hang for a few seconds as it tries to shove the larger photo data out the communications port. Also, exceptionally large or highly detailed photographs can be quite large, though it is rare that a webcam would create such a huge file. Just to be safe, check the size of the file and don't send anything larger than 5 MB if the receiving person has a standard 56K modem (or something slower) on the other end, otherwise it will take a very long time to get your photo. Theoretically, any size e-mail can get through, but proper Internet etiquette—sometimes called netiquette—requires that users warn recipients when large file arrivals are pending.

The other thing that you can do with a webcam and e-mail is to create video e-mail, which is basically sending captured streaming video to another user. This is a highly complicated process, however, and has a lot of constraints on it, such as available bandwidth and attachment formats. Because of this, I've decided to devote all of Chapter 9 to video e-mail. If video e-mail seems like something you would like to try, then that is the chapter for you.

The ultimate webcam setup is one that takes live, or nearly live, photographs and uploads them to a dynamic web page.

Basic web design

The ultimate webcam setup is one that takes live, or nearly live, photographs and uploads them to a dynamic web page. Some webcams also send live streaming video to the Web. The number of uses for this type of technology is nearly endless. Some day-care providers set up password-protected sites that let parents log on and view how their children are doing throughout the day. I've known pet owners that set up webcams in their house to see which animal is chewing the furniture. And businesses use webcams to give their shareholders access to daily updates at construction sites or to entice people to visit an interesting area such as a theme park.

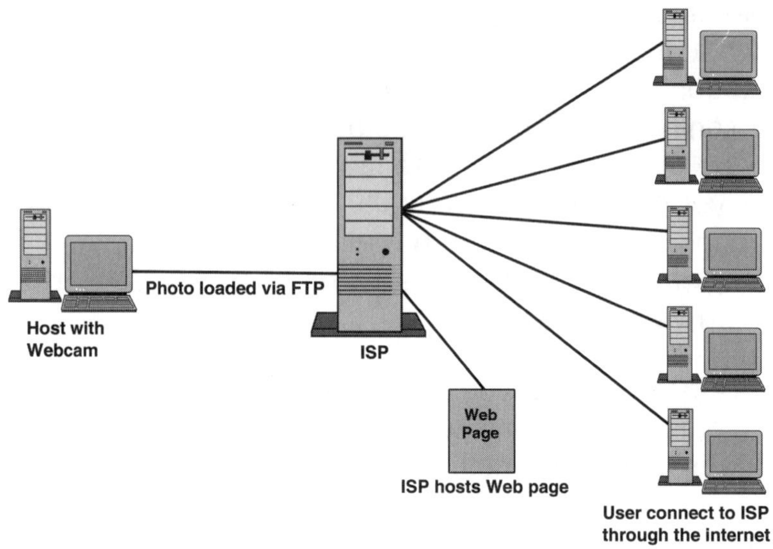

Figure 6-4. Diagram of FTP Upload setup.

You will have to check your software's manual or Readme file for specifics on how to set your operation in motion. But you will need a basic understating of the types of live webcam setups available and the requirements of each one. The easiest and most popular live webcam setup uses the FTP Upload method (see Figure 6-4). Cameras using this option take a picture every so often, and then send the image to the Web using FTP, or file transfer protocol.

The disadvantage to this method is that the images are not live. If you want to produce truly live images, you will have to use the streaming video camera setup, described later.

The main advantage to the FTP Upload method is that it can be done easily on a 56K modem. Sending a file using FTP is almost exactly like the way I described sending an image to a friend using e-mail. The only difference is that an image is automatically generated at set intervals, then sent to your Internet service provider (ISP) for distribution via your web page.

It would be helpful if you knew hypertext markup language, or HTML, which is the language all web pages are written in. If you don't, consider purchasing a third-party program such as Microsoft's FrontPage to make your web design easier. The reason you need to know HTML or how to operate a design program is because you will have to program your web page to accept new camera images when they're created by your camera software.

Here is, in a step-by-step process, how an image using the FTP Upload method gets to your web page. First, your webcam is filming an image. (Remember, the webcam is always watching.) You set your webcam software to take a picture of whatever the camera is facing every few seconds. I've seen this time interval as large as 10 minutes and as small as 10 seconds. On a 56K modem, you probably don't want it to be any less than 30 seconds between shots, because of the bandwidth constraints between you and your ISP.

Once the image is generated on your computer, your software takes that file, which is now saved in standard .jpg, .gif, or .tiff format, and uses FTP to send it to your ISP. The picture is then downloaded to your web page.

There are two methods of refreshing the webcam. First, you can do nothing. If you do this, the webcam will send a new image to your provider at intervals that you set up using your cam software.

There are two methods of refreshing the webcam. First, you can do nothing. The second method requires a third-party program called an applet.

The second method requires a third-party program called an applet. Certain applets can put more power in the hands of your viewers. They make it so whenever a user hits the refresh button on their browser, a new picture request is sent to the camera. This might sound really cool, but if you have many users they can quickly overload your cam site.

So, unless you have very few users (don't count on a good web site remaining secret forever) or a fast dedicated connection, stick with the timed uploads.

It is worth noting that even though this setup can be conduced easily with a standard 56K modem, you will need a dedicated "always-on" connection. Even providers who sell unlimited monthly 56K modem access to their sites don't normally allow users to stay connected 24 hours a day. Providers don't have one modem on their end for every user, because not everyone is logged on all the time. Many have five users for every one modem, or even 10 to 1 user-to-modem ratios. Normally, you can set up a dedicated always-connected account, but you will end up paying more in most cases, since the ISP must devote one modem for your exclusive use. Also, you really won't be able to do anything with the account except use it to update the camera, because the bandwidth needed to surf the Web would likely overload the camera's upload process.

Streaming live video is a bit harder to set up and has steeper requirements than the FTP-upload method.

Streaming live video is a bit harder to set up and has steeper requirements than the FTP Upload method (see Figure 6-5). Of course, your users are going to get live video, which in some cases is more desirable than just a series of still photos.

Due to the bandwidth requirements of a streaming video connection, you won't be able to do this on a 56K modem. Too many packets of information would get dropped due to latency on the Internet, and your viewers would be lucky to get blurry images and choppy sound. Mostly they would get a blank screen and no sound at all, especially when more than one user was trying to access your video feed.

To properly set up a streaming video camera, you will need at least a DSL (dedicated subscriber line), ISDN, or two-way

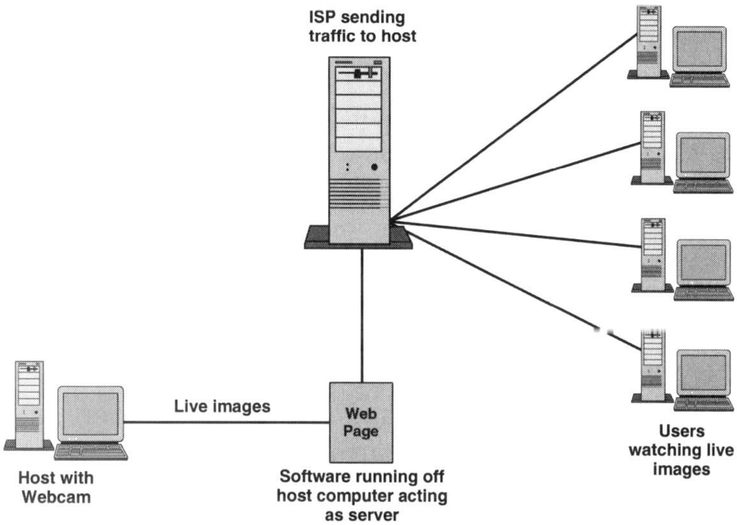

Figure 6-5. Diagram of streaming video setup.

cable-modem connection to the Internet. In the chapter on connectivity (Chapter 7), I'll discuss the pros and cons of each, as well as other methods. For now, just know that you need some kind of high-bandwidth dedicated connection, and that a 56K modem simply isn't going to meet your needs for streaming video.

When using a high-bandwidth connection to provide streaming real-time video, the video that users are seeing will come directly from your PC. Users still connect to the overall setup using your ISP, but when they're directed to your web page, they will see whatever the webcam is seeing at the time—minus a few seconds for latency over either the host or the receiver's connection.

How this works is that your web site will have to be programmed to accept data from your PC using an IP address and a port number. In this case, your web software is acting as your server, and your ISP is more or less acting as a match-up service directing users to the right place.

Even if you have the biggest bandwidth pipe money can buy, if you try to run your video full screen on your web page, very few users will be able to view it.

One thing to keep in mind when employing this method is to think about your users. Even if you have the biggest bandwidth pipe money can buy, if you try to run your video full screen on your web page, very few users will be able to view it. That's because the users will be trying to download all that data over their comparatively tiny 56K modems. Smaller bandwidth on the user side is not as much of a concern as small bandwidth from the host, but it is something to keep in mind.

When setting up your site, it's a good idea to keep the video window as small a percentage of the total screen as possible. Remember that each user will have a different size monitor running at one of a number of resolutions. So you can't always judge by how it looks on your setup or guess at how big the window will appear. You will need to measure the size of the window in pixels. The image from your cam can most likely be set to resolutions such as 640 x 480, 480 x 360, and 320 x 480 pixels. Most cameras support a maximum resolution of only 640 x 480. But for 30-fps (frame-per-second) video—or, as it is also called, "full-motion"—most cams will only support 320 x 240 with that refresh rate. Setting it to 640 x 480 will usually get you only around 8 to 10 frames per second. In addition, with anything larger than 320 x 240 you risk alienating users with dialup accounts and standard 56K modems.

Any software package that allows you to set up the type of webcam operations I just described will work just fine. But there are a few additional features that, while not necessary, are quite helpful. Depending on how you envision your cam setup, you should consider one or more of the following features when choosing webcam software.

Off-line image referral is one of those features that are almost essential, because no matter how redundant you make your setup, your webcam operation is going to be down at least once in a while. Off-line image referral lets you choose a set image or banner that is stored on the server. If for any reason the images stop updating, this feature automatically switches to the off-line image, which can be anything from a test pattern to an apology that the site is experiencing technical difficulty (see Figure 6-6).

Another helpful feature is the time and caption stamp. Depending on the type of webcam setup you are running, this could be considered almost an essential feature. The time stamp portion of this common feature simply puts the exact time within the photo that you are uploading, or in the corner of your streaming video. This serves as a reminder to users that the image is current, and it can also be used to archive important photos chronologically. If you're using this feature, it's important to find out where the software is getting the time information. Normally, it will come from either your host computer or off the computer that is hosting the web server. Wherever the information comes from, you will have to make sure the system clock on that computer is accurate. System

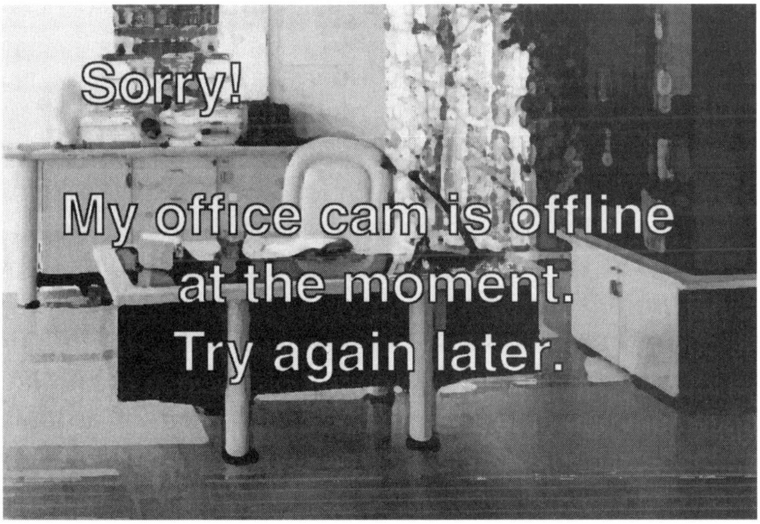

Figure 6-6. Photo of obvious off-line image referral.

clocks tend to lose seconds over time, so make sure you check the clock time at least every so often.

The second part of this feature, the caption portion, can be used to either explain to the viewers what they're seeing, or to insert copyright information into the image. Especially when you are running a business off the webcam site, it's good to make sure your images are legally protected.

When looking at features, don't take password protection for granted. Not all programs allow you to create a password-protected site. Even if you don't plan to use the password feature every day, having it can be helpful on special occasions, such as when you want to try something new with your images and limit access initially to a select group of critics or your staff of designers.

When looking at features, don't take password protection for granted. Not all programs allow you to create a password-protected site.

Some type of scheduler can also be helpful, especially if your webcam is not going to be on all the time. Perhaps you want to show pictures of your office from early morning to late afternoon, but want some privacy when working at night. With a scheduler, you can automatically set camera activation to occur during select hours and have it turn off after a certain time, otherwise you would have to do this manually every time.

The final feature that I would qualify as nearly essential is image-file numbering. This means that as the camera generates and saves files, it numbers them. If the first image is called something like pic001.jpg, then the 370th would be called pic370.jpg. At a glance, you can tell exactly which pictures were taken first and which ones came later. From an administrator's point of view, this is more helpful than a time stamp, because you don't have to open the image to know where in a sequence of photos a picture belongs. This feature is only useful if you plan to archive your photos, however, as it does not do much good to have a neat numerical system if you delete all your pics for space-saving reasons.

Of the three components to a webcam setup—the webcam, the computer, and the software—the software is the hardest to master. Most webcams are basically similar in design and close in image quality, and any computer that suits your needs will work as well as any other in terms of webcam operations. But software is more difficult to pin down. You can't really touch it, at least not the code itself, and the wrong package can quickly lead to a miserable experience.

So software is a bit like blood to your computer. The PC can't do anything without it flowing inside, but most of the time you don't think about it at all until something goes wrong and you're in desperate need of a Band-Aid. But don't fear finding and mastering your webcam software, especially if it is a free package or even shareware. Armed with the knowledge of the features you require and the type of setup you want to build, you can make an informed decision about what software you should purchase or download. And unlike an expensive computer, if you end up hating your software, it's not too difficult to shift gears and start over with a new package.

Of the three components to a webcam setup—the webcam, the computer, and the software—the software is the hardest to master.

The Connection: What a Tangled Web...

Chapter 7

OK, so you have your computer up to speed (or you have a new one), your camera is hooked up and ready to go, and you've figured out how to use the software you have. Now it's time for the big decision: what kind of connection are you going to use to connect to the Internet? Your choices are numerous, and among them you should be able to find something that will fit your varied needs (see Figure 7-1).

Now it's time for the big decision: what kind of connection are you going to use to connect to the Internet? Your choices are numerous.

The three main choices I'll tell you about are the traditional 56K dial-up connection, ADSL, and cable modems. I'll also be talking a little about ISDN, digital satellite options, and some of the higher-end business-related options. But my focus here is on the kind of connectivity a home or small office might require to fully take advantage of a webcam. If you are part of a midsize to large business, you'll want to talk to your

Figure 7-1. Diagram of various Internet connections. All use the same general building blocks, except for digital satellite Internet systems, which use different connections for receiving and sending data.

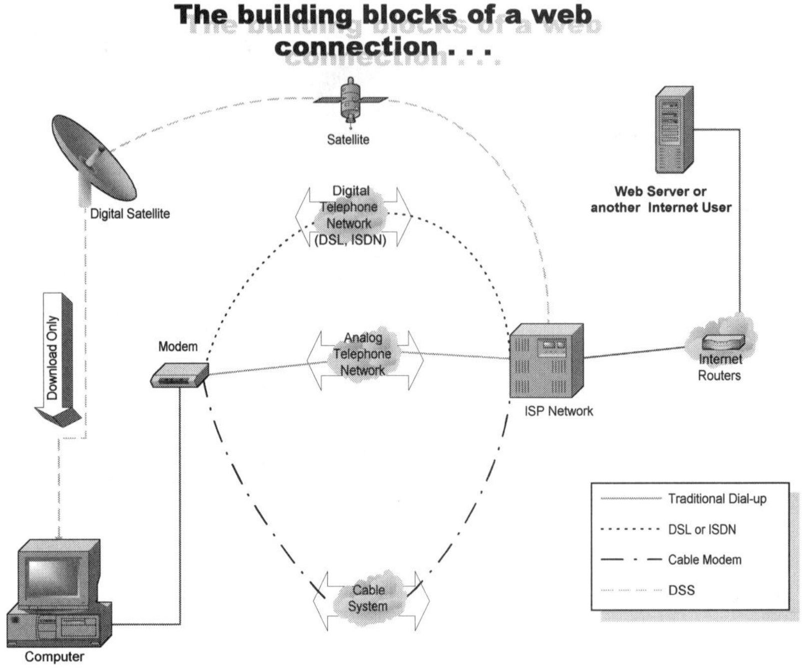

own technical support personnel to see what your connectivity options are and what you need to do to access them.

For the home user, the decision about which of these options to choose will revolve around a number of factors. Table 7-1 provides a checklist to help you decide which connectivity solution is best for you. Now I'll go into a little more depth explaining what all of this means and how it might affect your decision.

Availability

Obviously, with a 56K modem, your availability is wherever you have a phone connection. Right? Well, almost. Depending on the phone wiring in your house, your local area, and your phone

Connectivity Check List

- ***Availability***: Before you get your hopes up about how fast your connection can be, check to see the availability in your area. Many solutions, such as ISDN or ADSL, are only available in certain areas.
- ***Price***: This can often be difficult to figure out, as some connections come with some kind of rental equipment, usually a special type of modem. Do the best you can to figure out the cost of the equipment and service. Also be aware that many high-speed options don't include the cost of having an ISP, so figure on that additional cost.
- ***Speed***: How fast do you really need to go? This depends almost entirely on the kinds of applications you will be using. It is certainly comforting to overbuy here, and the feeling that you will essentially never run out of bandwidth is nice, but it is also very expensive. For some types of connections such as ADSL, you can actually buy more bandwidth later if you need to.
- ***Security***: Most people don't think about this issue in terms of their connection. For example, cable-modem systems are wired so that each connection is part of a local area network leading into the main cable system. What this means for you is that your neighbors and you could possibly see into each other's systems. There are steps to take to prevent this, which I'll go into later.
- ***Flexibility***: Sometimes you might want to connect more than one computer to a connection. Some newer operating systems allow you to share a connection, but this is a clumsy solution. It is usually far better to buy a small (eight-port or smaller) network hub and hook up the computers to that. Even with some connection options that support shared connections, you will have to check with the provider about whether they support that kind of usage.
- ***Usage***: What this basically means is: how long do you plan on being connected? Some connections are "always on," while others will disconnect you after a certain amount of time has passed and there has been no activity. If you are disconnected, you can usually just reconnect, but figure this into your plans when looking at your options.

Table 7-1. Connectivity Checklist to help you decide which connectivity solution is best for you.

provider's system, a 56K modem may not connect at nearly that speed. In fact, no 56K modem actually connects at 56K. The fastest speed possible is actually 53K, because of government regulations covering telephone communications. With older wiring, however, the fastest your modem may go might only be 28.8K. If this is the case, there's unfortunately not much you can do about it, as the phone company will either need to update its systems, or you will have to actually replace the phone wiring in your house. That would probably be a little much just to get a faster connection, especially considering that there is basically nothing wrong with your phone in terms of providing normal voice service.

One problem is that you often cannot find out if a modem will truly work at 56K before you buy it and try it out. Your local phone company can give you some idea about the wiring leading up to your house, but it usually cannot tell you about the wiring in your house. My advice is to either ask to borrow a friend's 56K modem and try it out, or purchase one from a store that has a good return policy.

One problem is that you often cannot find out if a modem will truly work at 56K before you buy it and try it out. Your local phone company can give you some idea about the wiring leading up to your house, but it usually cannot tell you about the wiring in your house.

If you really need a connection faster than 56K, or if a 56K modem doesn't really work with your current telephone system, then you will probably want to look at another option.

In some areas, mainly major cities and towns, your options might include something called ADSL, which is a specific type of technology broadly referred to as Dedicated Subscriber Line, or simply DSL. ADSL stands for Asymmetric Dedicated Subscriber Line, which uses a digital modem and your existing phone lines to provide connection speeds up to 126 times faster than a 56K modem. The problem is that the availability of ADSL is still very spotty throughout the country, as the technology is still being rolled out. Either your local phone company or a nationwide provider will have information about availability in your area.

ISDN is another type of connection, but one that seems to be falling out of favor of late. ISDN (Integrated Services Digital Network) has been around for quite some time. And though it seemed to finally hit it big in 1998, it is now seen as a less viable technology, due to the advent of ADSL and more mature cable-modem systems. Also, ISDN was a victim of its own success, with demand outstripping the ability of suppliers. IDSN, like ADSL, also uses your existing phone lines, but in many areas charged per minute rather than a flat fee. While ISDN is still available in some areas, I would recommend looking at other options.

So far, all the connectivity options I've discussed have revolved around either using your existing phone line or modifying it in some way. Now it's time to think outside the box, or more literally, *in* the box—the cable box, that is. Like ISDN, cable modems have been around for a while, though it is only recently that the systems have matured. Cable modems provide a network connection through your existing TV cable system. So availability is only limited by whether you have cable TV in your area and if your cable provider offers cable-modem service.

There is one other connectivity option that you might consider. The same digital satellite TV systems that can give you TV shows and movies can also give you the Web. There is a small caveat though: while the traffic from the Web comes via the satellite dish, the traffic from your computer to the Web actually travels by a standard 56K modem over phone lines. The advantage is that availability is only limited by whether you can get digital satellite service in your area, which most people can.

Price

Price is sometimes difficult to figure out. You'll often see specials for a specific price, only to find additional costs in the fine print.

As mentioned in the connectivity checklist, price is sometimes difficult to figure out. You'll often see specials for a specific price, only to find additional costs in the fine print. Be thorough in finding out and understanding exactly what is entailed with installing the service (e.g., any special equipment that is required) and what each item or service costs.

Most companies will provide you with an estimate if you wish, and it's always a good idea to outline the costs before committing to a service.

56K modems are by far the cheapest to use. Especially if you plan to use the modem with an existing phone line, there really is no new cost except for the cost of having an ISP to dial into. There are even options to have free 56K dial-in access, though you will have to put up with a lot of advertising (nothing's really free!). A good standard modem will cost anywhere from about $100 to $130, though a modem for a laptop will cost upwards of $190. The cost of an ISP dial-up varies on the whether you have a contract or simply pay month to month, but figure on around $20 a month. You will most likely be able to get it for around $14 or $15, if you commit to between a year and three years.

Of course, with solutions that offer higher speeds come higher costs. I'm obviously talking about DSL and cable modems. If you're looking for high-speed access, most of you will choose from between these two options. I really won't be discussing the pricing issues around ISDN, as the adoption of the technology seems to be in decline, and many companies have stopped offering it.

So how do DSL and cable modems compare on price? As always, there are layers to understanding the answer to that question, but in the simplest terms, cable modems are usually quite a bit cheaper. But there is price, and then there is value: what are you actually getting with each system?

If you're simply looking for the fastest and cheapest way to surf the Web, then cable modems are a good value.

If you're simply looking for the fastest and cheapest way to surf the Web, then cable modems are a good value. Once you start looking at providing content from your system to the Internet, however, there are problems with using cable modems. The first is that the cable company usually can't provide you with a fixed or static IP address. IP addresses are kind of like your Internet phone number, and they can either be constant or static (meaning your IP address is always the same), or they can be dynamic (meaning that it may change every time you connect to your ISP).

Without a static IP address, hosting your own server is impossible, and using programs such as Microsoft's NetMeeting (which uses your IP address as a way for people to find you) becomes more difficult.

A cable modem does not connect you directly to the cable company's network; instead, it routes you through a neighborhood network node and then on to its system.

Also, cable-modem systems provide one set amount of bandwidth, so you're either buying all or nothing. With DSL, most providers offer a spectrum of speeds. Often, additional ADSL lines can be bonded to an existing DSL line to provide even faster speeds, which is great for small or home-based businesses that need to be able to have their connectivity grow as they do.

With DSL, you're also buying security and reliable—sometimes even guaranteed—levels of bandwidth. Cable companies, on the other hand, usually cannot guarantee your speed, since they depend on how much usage there is in your neighborhood. This is because a cable modem does not connect you directly to the cable company's network; instead, it routes you through a neighborhood network node and then on to its system.

So while there is lots of potential bandwidth between you and the Internet, it all depends on that local network node, which you will probably be sharing with anywhere from 500 to 2,000 of your neighbors. Some cable companies may even have more users than that on a node temporarily. And suddenly all that incredible, advertised downstream speed doesn't seem so fast when split among you and 1,999 of your neighbors, does it? In fact, for short periods of time,

In understanding communication speeds, it also helps to understand that most of the options are asynchronous—that is, they transfer data faster in one direction than in the other.

congestion may be so bad that your downstream rates may actually be slower than a 56K modem. The upstream speeds, which are slower to start with, are also severely reduced when local congestion kicks in. Later, I'll talk about how all that neighborhood sharing not only reduces your effective bandwidth, but also increases your risk.

So I've established that DSL offers some advantages over cable modems, but what are the actual prices? It varies from region to region and provider to provider, but Table 7-2 provides some estimated prices to consider.

As you can see, the main difference in price is in the monthly fees. Special deals usually can be had, but usually they only reduce your one-time installation costs. So which is the best deal for you? Well, that will probably be decided by your feelings on the next topics: speed, security, and flexibility.

Speed

This is the area in which cable companies and DSL providers take advantage of the vagaries of technical specifications and fine print. Just as I discussed with 56K modems that don't actually transmit at 56Kbps, there are a lot of variables to consider when looking at what your actual data transfer speeds will be. In understanding communication speeds, it also helps to understand that most of the options we're talking about here are asynchronous—that is, they transfer data faster in one direction than in the other. This is OK, as most users receive far more data than they transmit when using the Internet.

The first most important thing to realize is that the businesses involved in on-line communications—whether manufacturers

Cable Modems vs. ADSL		
Item	Cable Modems	ADSL
Modem	$250 or $15 per month[1]	$250[2]
Installation	$99	$99
Monthly Fees	$40	$50 to $200[3]

[1] *Cable companies often require you to rent the cable modem for around $15 a month, rather than purchasing it outright.*
[2] *This is the retail price. Some DSL providers will only charge you as little as $99.*
[3] *Depending on bandwidth allotment.*

Table 7-2. The price of cable modems vs. the price of ADSL.

of modems, or cable and phone companies—mostly use the theoretical maximum speeds in advertising. So when you see someone advertising that their technology provides access speeds 25 times faster or even 126 times faster than a 56K modem, that's only the theoretical maximum and, in common usage, will never really happen.

The second most important thing to realize is that those speeds only apply to your connection from your home or business to your service provider's network. After that, your data is on the Internet, and the speeds your connection achieves are completely dependent on the amount of congestion on the path through which your data is traveling on the Internet, as well as the speed of the connection on whatever computer you're exchanging data with. So while you may have a super-fast connection, chances are that if you are trying to look at a web page of a company such as Microsoft in the middle of the business day, your actual speeds may not be much different than if you were using a

When you're talking about transmission speeds on the Internet, it is all very relative.

56K modem. Another good example revolving around webcams is using your digital video camera to chat with someone else. If the person on the other end has only a 56K modem (which is true of the majority of home users), then the speed and quality of your video and sound will be limited by his or her connection speed.

If you're still confused, let me put it this way: say you could write a handwritten letter to a friend four times a day, and after you finish each one, you immediately drop it in a mailbox. A mail carrier will come by that mailbox several times during the day and pick up your letters. The mail will still spend several days in transit, however, and the replies to your letters will take just as long to get to you. No matter how fast you get the letter into the mailbox, the overall "bandwidth" of your communication is essentially the same. Now substitute your computer for you, your high-speed connection for the mailbox, and the Internet for the postal system. Speed it up by several thousand factors, and you've pretty much described the state of web-based communications today.

The third most important thing to remember is that when you're talking about transmission speeds on the Internet, it is all very relative. For downloading simple text e-mail, you may save yourself a whole hundredth of a second by having a high-speed connection. Or you may save yourself several hours when downloading a large file. In terms of webcams, the main differences are seen when sending e-mail with video file attachments, when engaging in video chat, and when hosting live content for a web page.

Let's talk about how "fast" these high-speed connections really are and how you measure it. If you looked at Chapter 3, you've already seen the definition of "Kbps," "Mbps," and other such terms, but I'll go over them again just in case.

When people talk about speeds of transmissions between computers, they talk about bits per second. A bit, which stands for binary digit, is the smallest amount of information a computer can handle and is either a "1" or a "0." That means that one thousand bits equals a kilobit, right? Wrong. It actually takes 1,024 bits to make a kilobit, and 1,024 kilobits to equal a megabit.

Probably you also have heard of kilobytes, megabytes, and gigabytes and may be confused as to the difference. Well, a byte is simply 8 bits, just as a kilobyte is 8 kilobits. The reason for the difference is that the components within your computer—the memory, drives, and everything else—transfer data much faster than is possible outside of your computer.

To give you some perspective about what this all means, let's consider a normal 60-kilobyte file, maybe from a word-processing package. Your computer can open that file in a fraction of a second, but your 56K (kilobit per second) modem would need somewhere around 8 seconds to transmit that same file.

Usually, you will see these speeds of so many bits or bytes per second abbreviated rather than spelled out. The way to tell what someone is writing about is to look at the letter "b" of the abbreviation. If the letter is capitalized, as in KBps or MBps, then they are talking about bytes. If the letter is lowercase, as in bps or Kbps, then they are talking about bits. Strictly speaking, the first letter of the abbreviation should be lowercase, but in common usage these days you usually see it capitalized.

Probably you also have heard of kilobytes, megabytes, and gigabytes and may be confused as to the difference. Well, a byte is simply 8 bits, just as a kilobyte is 8 kilobits.

The speeds of your Internet connection are always given in either Kbps or Mbps, so we're looking at bits here. And the speeds of those connections can differ greatly, even across the same technology.

Table 7-3. The differences increased bandwidth can make.

Type of Usage	File Size	56K Modem	DSL - 384 Kbps	DSL - 640 Kbps	Cable and DSL- 1.5 Mbps
Browse four or five web sites	2.5Mb	6.25 minutes	53 seconds	34 seconds	13 seconds
Receive a short video e-mail	8Mb	20 minutes	2.75 minutes	1.7 minutes	43 seconds
Download the latest browser	25Mb	62 minutes	8.33 minutes	5.3 minutes	2.1 minutes

Cable modems are the hardest to pin down on the amount of actual bandwidth, as they are a combination of analog and digital technologies. Depending on the type of cable system installed in your area, download speeds may be as high as 1 to 3Mbps, while upload speeds will be as high as 500Kbps and 2.5Mbps. It is theoretically possible to get download speeds of 27Mbps and upload speeds of 10Mbps, but given the normal home-user setup, network congestion and everything else, those speeds simply aren't within reach for most users. Realistically, plan on around 1.5Mbps download and 200Kbps upload speeds. As cable companies update more of their infrastructure, speeds will increase, especially since providing Internet access will remain an extremely competitive market for years to come.

Bandwidth with ADSL usually can be guaranteed by the provider and offered at different levels, because it's a dedicated circuit. Generally, plans start at offering 384Kbps of bandwidth and go up to either 1.5Mbps or 8Mbps. Upstream speeds are usually not guaranteed, as they vary depending on how far you are from a central office. So speeds vary by system, area, technology, and seemingly even by what your neighbors are doing. Overall though, bandwidth is bandwidth and if you need it—there's no getting around it. Table 7-3 gives you an example of what kind of difference increased bandwidth can really make.

So is it worth it to spend the extra money for a faster connection? If you're going to use any of the webcam applications we just mentioned, or if you spend more than four hours a day in front of your computer, then the answer is probably yes. As I said before, DSL offers the best opportunity to buy only the bandwidth you need, but in the end it all depends on how you use your computer, the types of applications you run, and what other needs you may have.

Security

People seem to be both too paranoid and too careless in the use of the Internet, and connectivity is a key part of that.

Security over the Internet is likely to be the seminal computing issue for the first part of this new century. People seem to be both too paranoid and too careless in the use of the Internet, and connectivity is a key part of that. While you may choose a particular method of connecting to the Internet based on what your webcam needs are, keep in mind that security must also be part of the equation. Short of some type of wiretap, your 56K modem dial-up or your DSL line are physically secure from hackers. Yet the same cannot be said of cable modems, as the traffic in and out of your house can be passively "eavesdropped" by someone in your neighborhood.

That's why most cable companies have taken to encrypting or scrambling the signals between your cable modem and their system. This will keep out most people from discovering information such as passwords, credit card numbers, and even which sites you are browsing.

With computing power growing every year, however, cable companies will need to increase the complexity of the encryption of web traffic across their systems to ensure they stay secure.

While almost all cable companies should have some sort of encryption in place by now, make sure you double-check with your cable provider if you are considering cable-based Internet access.

Chapter 15 covers computer security from top to bottom. I haven't discussed the more esoteric ways in which the data you send or receive from the Web can be "poached" from your connection. Most require very technical and expensive equipment, and the simple truth is that no matter how secure or insecure your connection, your data is still going out into a public area where it can be looked at. The trick is not to look at security as an issue revolving around a single piece of your computer setup, but to look at it in a more three-dimensional or holistic fashion. And that's what Chapter 15 will hopefully help you do.

The trick is not to look at security as an issue revolving around a single piece of your computer setup, but to look at it in a more three-dimensional or holistic fashion.

Flexibility

When I talk about flexibility in your Internet connection, I'm not talking about how much you can bend a cable. No, the idea here is to consider in what ways your connectivity option can change or grow with your needs. After all, most of the high-speed bandwidth options come with some hefty installation charges, and you don't want to discover at the last minute that you're going to have to change horses, as it were. To highlight what you'll need to consider, take a look at Table 7-4.

When looking at the number of users a particular type of connection can support, the question is actually two-fold. First, you figure out how many people in your house or business are going to be using the connection, and then you need to count how many computers are being used. After all, it's not unusual to have children sharing a computer and Internet access in a home, for example. The reason you count them

What do I need my connection to do for me?
• How many users do I need to support? • How many computers do I need to have simultaneous access to the Internet? Does the connection provider support multiple computers? • Do I need a static "fixed" IP address, or can I use a dynamically assigned IP address? • How easy is it to increase my bandwidth? • How stable is the technology used, and will it be around for three years? Five years? Ten years? • Do I have to use the connection provider (or a company of their choice) as an ISP, or can I use my existing ISP? • How easy is it to move the connection's entry point into my home in case I need to move my computer? • Is a service agreement included, and does it cost extra? What does it include? • Is there a guaranteed level of bandwidth? Does it only cover bandwidth in one direction (upstream, downstream)?

Table 7-4. Flexibility checklist. Consider what you need your Internet connection to do for you.

separately is that, in those cases where the company providing your Internet connection is also the ISP providing you with e-mail boxes, you'll need to figure out how many you will need. If you need more e-mail addresses than your ISP provides, you can get around that limitation.

There are, for instance, many options for getting free e-mail addresses through such services as Microsoft's Hotmail or Yahoo! And the reason you need to consider how many computers you may have connecting to the Internet at any one time is that you'll need unique IP addresses for those computers. Even 56K modem users have to deal with this. If you have two phone lines and two computers, you can often connect both at the same time, but you may incur additional hourly charges—even if you have a so-called unlimited-access account. Normally, with DSL and cable modems, they simply will charge you around $10 a month for an extra IP address. So if you managed to get cable-modem access for $30 a month, and then add $15 for the rental of your cable modem, you will still have to pay $10 for each additional computer you intend to have connected at the same time.

If you have two computers and want both of them to be able to connect to the Internet but not at the same time, then you don't have to request additional IP addresses and won't be charged extra. To change which one is connected, however, you may have to completely shut off the computer that's currently connected and wait a certain amount of time for the modem to reset, before attempting to connect with the other computer. This is usually the case with cable modems. With DSL modems, usually you only have to choose to disconnect using your software and then connect with the other computer. Even with DSL, however, you may have to wait a few minutes for the modem to reset.

As I said before, you will need to get extra IP addresses if you plan to have more than one computer. But even if you only have one, pay attention to what kind of IP address your service provider has. If you plan on hosting any kind of information from you computer to the Web, or if you need to have a fixed address for such things as video chat, then your options are limited.

You will need to get extra IP addresses if you plan to have more than one computer. But even if you only have one, pay attention to what kind of IP address your service provider has.

A growing number of cable companies and DSL providers are relying on dynamic IP addresses. This allows them to support more users on fewer addresses, since everyone is usually not connected at the same time. This has been common practice for 56K dial-up ISPs, and now the other ISPs are following suit. Since these connections are "on" or "live" all the time, you may wonder what benefit they derive from this, as they have to have enough IP addresses for everyone.

Well, I've never had it confirmed in person, but you have to suppose that ISPs want to limit how much traffic is going across their system. Having a dedicated IP address usually means that you can host a web site or easily use video teleconferencing and other high-bandwidth applications. If they require you to use a dynamic address, then they can stop you from using the types of high-bandwidth applications that require a fixed IP address.

The issue of bandwidth is central to connectivity, but the different technologies don't all handle it in the same manner.

If you actually need a static IP address to host a web server, for instance, then you might have to look at other, more expensive options (such as a T1 line), or look for a company to host your server from its own premises. Also be aware that many providers forbid users from hosting web sites through their service. Some only forbid business web sites, while others have a wholesale ban on hosting any pages.

The issue of bandwidth is central to connectivity, but the different technologies don't all handle it in the same manner. As I briefly mentioned before, depending on the provider and the type of lines into your home, DSL can offer you more bandwidth on your existing line or add another DSL line that can be "bonded" with your existing one to create an even faster connection. It is theoretically possible to do the same with cable modems, but the benefits would be so low (because of the shared nature of the bandwidth) that there really would be no point. So in terms of bandwidth management, DSL offers much more flexibility.

The life cycle of the technology you're using also should be considered when looking at the flexibility of the solution. If the technology is going to change in three years, then you might be left out in the cold or forced to migrate to a different solution that doesn't meet your needs as well. The good news is that the technology standards being used for both cable modems and DSL are designed to evolve, and while new standards are constantly being developed, they will more than likely be backward compatible.

When looking at the flexibility of any Internet connectivity solution, you must consider where your connection is to be placed physically.

The most likely switch you would have to make (if one was required) would be to change your modem. Since cable companies usually rent their modems, they will most likely provide a replacement. DSL modems are usually purchased, however, and a drastic enough change in the technology might require you to buy a new modem. This really hasn't become a huge issue yet, but in the next five years it may become more relevant.

When considering Internet connectivity options based on your webcam needs, you may consider upgrading from one type of service to another. In some cases, you can maintain your current ISP account and just change how you connect. Many dial-up ISPs are partnering with DSL providers, giving their customers an easy upgrade path. Changing the method you use to connect will usually require you to switch ISPs though. So count on having a new e-mail address and letting people know about it.

Also, you will probably want to back up and save your e-mails from your old account. Usually the best advice is to run both services for one month concurrently. That eases problems and makes the transition much easier.

As outlined in the flexibility checklist, there are three more areas to consider when looking at the flexibility of any Internet connectivity solution. You must consider where your connection is to be placed physically. If you're using a cable modem, for example, you might need to have the cable company come and move your cable connection if you ever need to move your computer. The same is true of DSL, in that an additional phone outlet may need to be installed if your computer needs to be moved. Of course, with a 56K dial-up connection, you need to have a phone outlet handy, but often you can do that yourself. For DSL and cable-modem customers, a more flexible approach might be to connect your DSL or cable modem to a network hub in a central location, and then use a network cable to connect to the hub.

With many homes now having more than one computer, this not only provides you with more leeway when considering the physical location of your computer, it also allows you to network your computers together. Another bonus is that, if you have one or more digital video cameras that connect directly to a network connection, you can simply plug them into your existing hub.

As you may have guessed, since these connectivity technologies are still growing and evolving, technical support and a service agreement can be important parts of your webcam experience. The first thing to determine is if your ISP provides a service agreement, and then you'll need to find out exactly what is covered. Have in mind how you want to set up your computer and webcam, and talk with your ISP about it before you sign up. It's far better to determine beforehand exactly under what circumstances you can get support and what it will entail.

Guaranteed bandwidth is basically an assurance from your provider about the levels of bandwidth that under any circumstances you will have. This is not dependent on the technology, but rather on the provider.

Lastly, a subject I touched on briefly before is guaranteed bandwidth. This is basically an assurance from your provider about the levels of bandwidth that under any circumstances you will have. This is not dependent on the technology, but rather on the provider. I have heard of both DSL and cable companies providing guarantees and also have heard of many that don't. Depending on what your plans are for your webcam and other bandwidth-impacting activities, this may or may not be a concern. Generally speaking, having a guaranteed amount of bandwidth is always preferred. Yet, in and of itself, it's not enough to make you choose whether to go with a particular service if it meets your needs in other areas.

Usage

The last subject on the connectivity checklist is usage. By this I mean restrictions the provider puts on your connection. These can be either restrictions on the amount of time a connection can be used or on the amount of bandwidth your connection generates. For years, dial-up providers have disconnected users whose connection has not been active for a certain amount of time, usually somewhere around 20 minutes. Also, some dial-up providers still offer limited-use accounts that are cheaper, but only give you so many hours per month of connection time.

DSL and cable modems are designed to be on all the time, and DSL providers usually place no restrictions on the amount of bandwidth you can use. Some cable companies, however, still have bandwidth ceilings on their cable-modem subscribers. If this is the case with your local provider, check to see how much it is and if you will likely reach that limit during a month. If it seems you might, you will probably want to look at other options or reexamine how you plan to use your Internet access.

DSL and cable modems are designed to be on all the time, and DSL providers usually place no restrictions on the amount of bandwidth you can use.

Summary

As you can see, the type of connection you have plays a vital part in determining how you use your webcam and vice versa. Also, individual providers often have unique requirements and restrictions, either based on the type of technology they are providing or on their business models. The most important thing is not to feel pressured to subscribe to a particular provider just to get higher-speed access. As with all things computer-related, six months down the road things are often cheaper and easier to use. Balance your needs and wants against what is available. If you can't make a real case to pay the higher prices for more bandwidth, then simply wait until you can. Access will get cheaper and easier to come by, and this is a trend that will continue for the foreseeable future.

Webcams are wonderful tools, but they need fast, reliable Internet access for you to use them to their full potential. The bonus is that faster and more reliable access can be a boon in almost every other application on your computer. From faster downloads to better communication and beyond, high-speed access is the direction we're all headed in. Whether you get there first or last is not important, as long as you have the right level of access for your needs.

As with all things computer-related, six months down the road things are often cheaper and easier to use.

PART 3:

Getting Started With Applications

Capturing Video: Hunting For the Perfect Picture

Chapter 8

Recording your own videos is very much like taking photographs, but with a few differences. The art of "filming" with a webcam is one that is easy to learn, but much more difficult to master. Since a webcam loads its images directly into a computer, there are some special considerations (such as dealing with movement) that a person with a video camera will not have to worry about. But generally, using a webcam is much easier than learning to use a 35mm or a VCR camera.

There are four basic elements to creating a good video using a webcam.

There are four basic elements to creating a good video using a webcam. If you pay attention to lighting, composition, and continue to practice your craft, you will be creating award-winning videos in no time. Whether you want to send a movie of your new puppy to your friends or a business proposal to a client, your video will appear well made and professional.

There are two main enemies to successful video-making through a webcam: light coming in from the wrong angle, and not enough light to begin with.

Working with lighting

With a 35mm camera, working with lighting is a complicated process involving calculating the focal length of your lens versus the shutter speed in the given lighting conditions. Master photographers can almost calculate this value in their heads, and there are many handheld devices that help by measuring the ambient light. Plus, a photographer can cheat somewhat by using a flash to fill in the shadowed areas, even during daylight photography. When I was a newspaper photographer, capturing the depth of field in most pictures—such as a raging fire or a burning car wreck—became the hardest part of the job. After two years in the trenches, however, I could do it with ease.

Luckily for webcam users, working with lightning is not nearly so difficult. Webcams basically use the same technology as a VCR camera, in that they simply "film" everything in front of the lens. As such, providing enough lighting, and the right type, is far easier to do—although this ease comes with some added worries.

The main problem when working with lighting is that the camera has no light source of its own. It basically sits there and records whatever comes into its view. So there are two main enemies to successful video-making through a webcam: light coming in from the wrong angle, and not enough light to begin with.

Let's start with the problem of not enough light. This is a common problem with any type of digital image-capture device, as they all perform somewhat poorly under low-light situations. It would seem that just adding more light would fix this problem, but in fact, if the wrong type of light is added or if light is positioned in the wrong location, more light can actually make the problem worse.

Almost every office in the world—and even the occasional room in your home—is lit by fluorescent lights. When you try to photograph someone using a 35mm camera without the benefit of a flash, the person or object will end up having a greenish tint. This is because fluorescent lights actually produce a light that is slightly toward the indigo range of the spectrum. Also, fluorescent lights actually flicker—too fast for the naked eye to see—and fast camera shutter speeds can actually be unlucky enough to snap the picture during the dark part of the flicker. This means the picture will turn out dark, not to mention having a green tint.

Webcams are basically immune to the flicker problem, because they watch the world much like your naked eye. Even when taking a picture, they can't stop the action enough for it to matter. But you will suffer from two major problems when working with fluorescent lights. First, you will suffer from the green tint problem. This is unavoidable, although it will be much less drastic with a webcam than it would be with a 35mm camera.

Since the webcam does not have the advantage of a flash to fill in harsh shadows, even in very bright top-down light you are not going to be able to capture a lot of detail off of a person's face.

The other problem with that type of lighting setup is also the second major problem you will encounter overall, even in most situations when the light is not fluorescent. The problem is the angle that the light comes onto the subject. Most fluorescent lights, and many normal bulbs in private homes, are mounted in the ceiling.

Since the webcam does not have the advantage of a flash to fill in harsh shadows, even in very bright top-down light you are not going to be able to capture a lot of detail off of a person's face. Depending on how your subject styles his or her hair and the amount of wrinkles in the face, you are going to generate a lot of shadows using any light that's positioned above where the camera is recording.

The obvious solution is to place some type of light source somewhere behind your webcam. This can be as simple as moving a lamp to a position behind the camera, or it can be much more elaborate. The light source does not have to be directly behind the camera, but it should be behind the invisible plane that extends across the front of the webcam's lens.

For all its high technology, a webcam is still a fairly primitive camera. If a user does not pay close attention to the lighting, video from a webcam won't look anywhere near the professional nature viewers have come to expect from television and movies.

This fact became painfully evident at a recent trade show I attended that, for some strange reason, insisted on having each attendee wear a badge with his or her picture imprinted on it. This was done by taking a snapshot of a group of people using a webcam and then printing out the photographs on a photo-quality printer. The photographs were then laminated onto the badge.

For all its high technology, a webcam is still a fairly primitive camera.

Now, forget that this was not a technical-type convention—in fact, it was a science-fiction convention. In any case, this kind of use of a webcam is not recommended. The room where registration was taking place was illuminated by fluorescent lights that were positioned around the top of the room near a sky light. It was nighttime, so there wasn't ambient sunlight, and I believe there wasn't even a moon that evening. To make matters worse, the hotel had a bed of ferns all along the top of the wall in front of the lights, blocking to some degree what light was filtering down.

I knew we were in trouble when I saw that there was a mob of increasingly angry people packed into the room. Doing what I normally do, I stood back quietly and took stock of the situation. It seems the convention organizers were using the webcam to photograph people in groups of four or six. The problem was that when the photos came out on the other end, the people all looked blue. Worse yet, their faces were so dark you might as well have pasted a blue splotch on your badge to represent yourself.

If a user does not pay close attention to the lighting, video from a webcam won't look anywhere near the professional nature viewers have come to expect from television and movies.

The organizers had apparently trained their security officers to scan the photos on the badges, a task I assume let them save some money by creating cheap and easily reproducible badges, but backed by a nonreproducible photograph. They didn't want to forgo the photograph for fear of counterfeiting, but the equipment was not cooperating.

As sometimes happens in these situations, the call went out for a techie to fix the problem. It was the proverbial "Is there a doctor in the house?" hue and cry. Amazingly, in a room full of sci-fi enthusiasts, nobody came forward.

Seeing that this was a webcam problem, I decided I might as well give it a look. (The Japanese animation marathon was getting ready to start, and I didn't want to be stuck in registration during the screening of *Vampire Hunter D*.)

Scanning the software, I discovered that the camera and the rest of the system were, for the most part, configured correctly. The problem was lighting. There was enough light in the room—just barely—but it was all coming from the ceiling. The camera could not accurately capture the details in people's faces, a must for security identification.

Thinking quickly, I acquired a little decorative lamp from the hotel that sat about a foot off the ground, which incidentally was exactly the height the webcam was on the table. I placed the lamp behind the camera and off to the left side, so people getting their photos taken would not have to stare at the naked bulb. That was better, but still not good enough. Webcams can be tricky, and light is not something you want to mess with if you need a quality image. I commandeered a silver fruit bowl from the hotel, some duct tape, and a ream of aluminum foil (taken off a really poor robotic costume). Lining the bowl with the foil, I secured it in place behind the lamp with the tape and faced the bowl outward. The result was a beam of yellowish light directed straight where the photos would be taken. This time the photographs came out great, with every painted cheek coming out beautifully on the other end.

When I got back home, I retreated into my lab to experiment with webcams in different lighting conditions. I have come up with an ideal lighting setup that should be mimicked, in whole or in part, as much as possible (see Figure 8-1). You don't have to be too elaborate; after all, the convention worked just fine with a fruit bowl, duct tape, some foil, and a decorative lamp. But this will give you an ideal situation to try to attain.

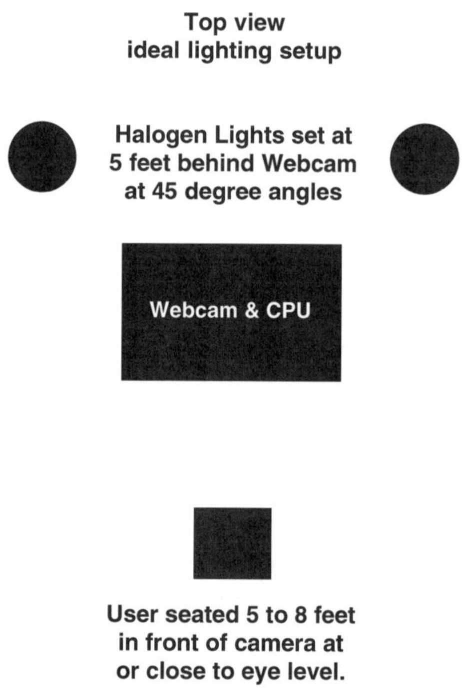

Figure 8-1. Diagram of proper lighting setup that you should try to mimick as much as possible.

First, the webcam should be positioned about five to eight feet from the subject being filmed and should be on the subject's eye level or slightly below. Second, two halogen lamps of about 500 watts should be set about five feet behind the camera at 45-degree angles. Mathematically speaking, to achieve the optimal angle of light diffusion (assuming the lamps point upward like they normally do), the top of the lamp should be no more than three feet above the top of the lens on the webcam. If you set up a room this way, the height of the lamps should not be a problem, given the average height of desks and chairs.

The bulbs don't have to be set to maximum power, and you can use fluorescent or other overhead lightning to help throw more light into your subject's area if you choose. You can even use normal light bulbs, though I prefer halogens because they're more flattering to skin tones, offering a richer, more detailed look. It's the closest thing to sunlight you can reproduce yourself.

Beware of backlight. Webcams are very much influenced by backlight due to the way they process images.

Finally, beware of backlight. Webcams are very much influenced by backlight due to the way they process images. If you have a very bright light sitting behind a person, such as a mini-fluorescent bulb from a cubicle, the webcam will more often than not try to compensate for the bright light and will make the entire image dimmer. This means that your subject, which is the entire focus of your video, will be drowned in artificial shadows. So make sure all your light sources come from the front of the area you are recording, behind the lens of the camera, and are of a type that's flattering to your subject.

In addition, I have a small caveat for using a webcam in the presence of natural sunlight (either indoors or outdoors). While this will often give you a more "natural" look, make sure that the lens of the webcam is never aimed directly at the sun for any real amount of time. Most manufacturers will warn you of this in the manual. Due to the brightness of direct sunlight and the focal points of the lenses, you might damage the CCD or "electronic eye" of the cam, so be aware of this.

Even with perfect lighting, the wrong composition can confuse your intended viewers, shifting the subject away from the proper target.

Composition is key

Whether you're taking a still web photo or making an extended video, composition is the second key to a professional look. It's also one of the hardest items to master. Even with perfect lighting, the wrong composition can confuse your intended viewers, shifting the subject away from the proper target. At worst, the video or picture can appear awkward.

One of the trade secrets of professional photographers, and one that translates well to the world of webcams, is the rule of threes (see Figure 8-2). Sometimes called the golden mean or the grid pattern, the rule of threes is the same no matter what name it goes under.

To practice the rule of threes, you simply divide the area you are about to photograph or film into a grid pattern. This is exceptionally easy to do using a webcam, because for the most part, the camera doesn't move. Also, you can watch the area that is about to be filmed right on your computer screen and even position yourself or the subject on the fly.

Once you have the grid pattern in your mind, laid out like a tic-tac-toe board, the next trick is to put your subject in the correct place. The optimal placement for a subject is exactly where the lines intersect. Suitable placement is also anywhere along the grid lines, and in some cases is actually better than an exact intersection.

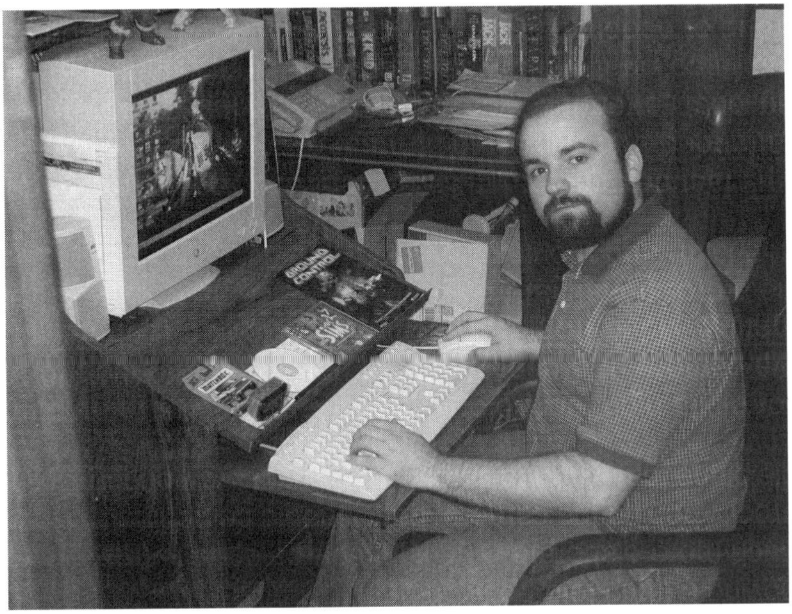

Figure 8-2. Photo properly demonstrating the rule of threes.

Take a look at any number of news photographs and famous museum-type pictures and you will find that, almost universally, they follow the rule of threes. When you watch movies or even the evening news, you will notice the same thing, once you know what you are viewing. Take the evening news as an example. You rarely have an announcer in the dead center of the screen. More often, you have two anchors, seated along the grid lines beside one another.

Your webcam videos can follow this same pattern, especially if you want to have two people in your video. Simply sit each one along the vertical grid pattern. The video will appear professional and pleasing to the eye, even if the viewers don't know why.

Single-subject videos, such as when only one person is speaking, are a bit harder to set up. At times, putting someone along a grid line will look odd, depending on the background. The easy way to compensate is to have something balance your subject, like having him or her sit sideways and giving the subject papers or charts to hold. You can also balance it somewhat by zooming in, if your camera is capable. And even if it's not, you can always sit the subject closer to the camera or move the camera closer to the subject. This will decrease the space between grin lines so that the subject is only slightly off center.

Another concept of composition is framing. This is somewhat difficult to do with a webcam, but not impossible. When taking a photograph, framing means shooting through an object so that it appears to frame the subject.

The finest example of this I have been able to produce was when I was taking photographs of Washington, D.C., in the springtime for a local magazine. Shooting the monuments is a relatively easy task, but it has also been done millions of times. So what I did was to photograph the Jefferson Memorial through a blossoming grove of cherry trees. The monument was the subject of the photo, but the leaves and blossoms of the pink tree made a slight frame in the upper-left corner of the photo. Not only did this capture the feeling of what I consider one of the world's most beautiful cities in the springtime, but also it was like taking a photograph in a ready-made frame.

The framing principal does not have to be taken literally. You can have an object just touching the corner of the photograph or running down one side.

The framing principal does not have to be taken literally. You can have an object just touching the corner of the photograph or running down one side. In fact, if you try to make the frame go all the way around your image, it won't look right at all. For a webcam, try sitting an office plant up near the front of the lens or the corner of a laptop computer (see Figure 8-3). It should be just enough to give your audience a sense of depth.

One cautionary note when using framing with an auto-focusing webcam: be careful that the object you are using to frame

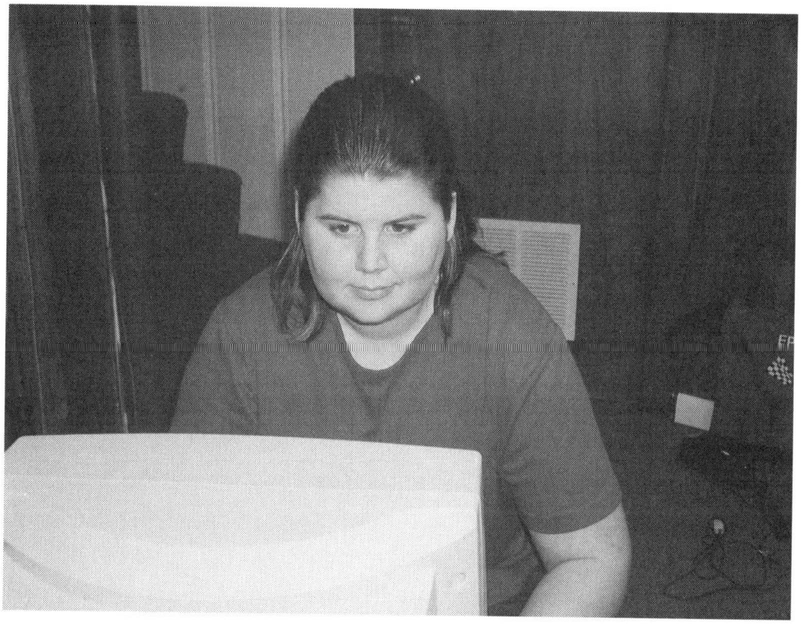

Figure 8-3. Example of framing in a web conference.

is far enough outside of the center of the lens so as not to trip the focusing program. If you stick a leaf a third of the way inside the view, chances are the camera will detect it and try to focus on the closer object instead of the actual subject.

On some cameras I have experimented with, it was impossible to do any framing at all, because the auto-focus on the camera was too sensitive. The camera either focused on whatever object I tried to use, or kept bouncing back between keeping the subject and the framing object in focus.

This focusing problem turned into an embarrassing situation during a conference with some editors in California. I had set the camera up with a jar of pencils to be used as my framing object. Everything looked fine, but as the conference began, the camera suddenly started focusing somewhere between the pencils and me. The result was that suddenly, nothing was in focus. I played with the controls and tried to pretend nothing was happening, as my editors talked to me about my latest assignment.

Finally realizing what the camera was doing, I subtly moved my hand under the focus range of the camera and flicked the jar across the room in one swoop. A few seconds later, the camera had refocused on me.

So, with some equipment you won't be able to frame. But unlike the rule of threes, framing is not mandatory; it's just a nice touch if you can figure out a way to make it work.

The reason you want your webcam shots and your videos to be simple is so your subject doesn't get lost.

Keep it simple

When I tell people about the need to keep things simple, they quickly begin to ignore me. Apparently, making something simple is equated with being easy. But tell that to a scientist trying to reduce a complicated equation. Sometimes taking something that is complex and making it simple is no easy task.

The reason you want your webcam shots and your videos to be simple is so your subject doesn't get lost. If you've ever watched a television newscaster trying to give a report from a public event when tons of idiotic spectators are blowing kisses and giving high fives and making various other gestures in the background, then you know what I mean. Nobody ends up looking or even listening to the reporter when there are so many more members of an animated, drunken hoard patrolling the background of the shot.

Now, hopefully you don't have any animated, drunken hoards in your home or office, but there are myriad distractions that can plague a webcam video or take away from the subject of a photo.

Imagine what the webcam is about to shoot as a painter's canvas. To begin with, you have a fairly broad subject, which as you learned earlier is everything that happens to wander in front of the unblinking eye of the webcam. Chances are, there is going to be at least one distracting element somewhere in your shot.

You can eliminate these distractions in one of two ways. First, you can physically move the distracting object out of the shot. Deciding what might be a distracting object is often more difficult than moving it. Once, I was watching a web broadcast following the release of a new type of notebook computer. The president of the company in question was talking about how the new notebook would revolutionize mobile computing. But he had three model cars on his desk during his presentation. I can't remember a thing he said, because I was too busy examining the cars and trying to figure out if they were Matchbox or Hot Wheels vehicles. (I know it's silly, but my father collects Matchbox cars, and one of the vehicles on the desk I had not seen before.) The point is anything can be a potential distraction. If it's a unique object and it has nothing to do with what you're talking about, get it out of your shot.

Imagine what the webcam is about to shoot as a painter's canvas. To begin with, you have a fairly broad subject, which as you learned earlier is everything that happens to wander in front of the unblinking eye of the webcam.

In another instance, I was talking over a webcam link with a public-relations manager working out of her home. Suddenly, the fattest cat I have ever seen in my life jumped onto the bookshelf behind her. I just kept thinking, "My God, that is a fat cat" the whole time she was trying to sell me on her company's product. To make matters worse, the cat looked like it was contemplating leaping down on the poor woman's head, a feat I thought might kill her. She must have noticed me starting to laugh, because she excused herself to take the hissing feline out of the room. In the distance I could hear the cat screeching with displeasure and her yelling at it, since she was still wearing her radio mike. The entire thing was so funny; to this day, I can't remember what she was selling. So do yourself a favor and take all the toys out of your filming space and lock humongous cats in the closet for a little while.

The only problem comes when you can't physically move an object out of your shot. For example, if the executive in the first example had an entire wall full of toy cars behind him and he absolutely had to shoot his part of the webcast from that location, it would not be very easy to move the entire shelf.

So then the second solution to eliminating distractions is simply to bring the focus tighter on the subject (see Figure 8-4). This, in effect, crops the photograph down so that distracting elements are not included in the shot, though I would advise moving to a less busy background. Most webcams will let

Figure 8-4. Example of a very tightly cropped photo, eliminating distracting elements.

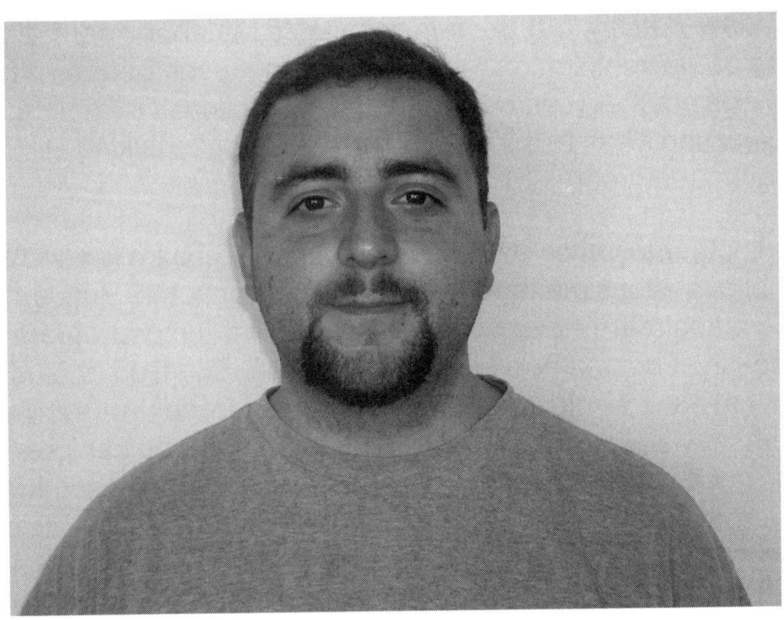

you crop your photos or videos down to a space about three feet in front of your subject, though I have also seen many cameras that let you literally stick your eye up to the lens and have it displayed perfectly. The latter type of camera will let you crop your photos and videos as much as needed.

Make streaming videos of yourself to practice for when you will do the real thing, then e-mail them to yourself or a friend and sit back and critique your work.

Practice makes perfect

Although film-type camera photographers have to generally spend a lot of money honing their craft (or find an unsuspecting newspaper to buy all their equipment for them), webcam photographers don't have to spend hardly anything except for the camera itself. There is no film involved and no excuse for not practicing.

A less well-known secret in the photography community is how many photos actually end up on the cutting room floor. For every photograph you see in a magazine of a fireman carrying a baby out of a raging fire, there are probably hundreds of shots taken that day that did not come out, or were not up to the standards of the publication or the photographer. When I used to run accident and fire calls with my camera for a small-town newspaper, I carried at least five rolls of film with 24 exposures. Thankfully, the newspaper was buying all my film or else I would not have had that luxury.

But with a webcam, it's all digital ink, so to speak. If you don't like a shot, simply delete it. Make streaming videos of yourself to practice for when you will do the real thing, then e-mail them to yourself or a friend and sit back and critique your work. The ins and outs of this process will be discussed in Chapter 9. I often find that I am my harshest critic, though any mistakes I make in practice runs can be eliminated when it counts.

The formula for capturing motion with a webcam is to add one part art, one part science, and top it all off liberally with a dose of luck.

In short, practicing with your webcam costs nothing other than your time. And time well spent in practice can cash out to be worth a lot more if you can avoid embarrassing and amateurish mistakes when trying to make a professional photograph or video. As with almost anything else in the world, practice makes perfect.

Freeze!

Some people just can't sit still, and if you're unlucky enough either to be one of these people or to be filming one, you may be in for a bit of trouble. Recording video to a computer's hard drive is a lot different from recording to videotape.

When you record to videotape, the tape is more than sufficient to handle all the data that is being thrown onto it. Some fancy cameras can even speed up the rate at which the tape is being put through the camera, resulting in slow motion when the tape is played back at normal speed.

But you won't have as many options with a standard webcam. The formula for capturing motion with a webcam is to add one part art, one part science, and top it all off liberally with a dose of luck. Also, individual cameras perform quite differently, so it's best if you have practiced often so you will know what your individual camera can and can't do.

Webcams are also greatly affected by the type of hardware that is driving them. You can easily get perfect results with a camera on one computer, only to have it fail miserably when you attach it to another.

The rule of thumb when capturing video with a webcam is to avoid as much movement as possible. What happens is, since a webcam must rely on a video card to record motion, often the refresh rate is not set high enough. And the refresh rate can only be set so high before it starts to overload the processing cycles of the computer. The result is that movement often is recorded very poorly by the camera, leading to blurry motion lines.

Typically, this means that the subject will remain in focus while the background is largely ignored by the camera. The software is often smart enough to concentrate the recording power of the camera on the subject, since even a person sitting still will have slight movements. But, if the subject moves his hands in an arc over his head, what will happen is that the camera will sometimes be slow to record that motion. Often the moving hands will be recorded, but because they passed over the mostly ignored background, the background will be blurred.

Once the person returns to a normal sitting position, it will appear as if parts of the hands are still over the head. The camera will continue to concentrate on the main subject of the video, and will only slowly update the background as processing cycles become available. Sometimes this blurry background effect will continue for several minutes.

As long as the subject does not do a lot of "talking with the hands," even a low-end camera on a cheap computer will be adequate for most presentations.

The best way to avoid this problem is simply to reduce movement as much as possible. As long as the subject does not do a lot of "talking with the hands," even a low-end camera on a cheap computer will be adequate for most presentations. Of course, if you're filming a sign-language interpreter or have a presentation with a lot of charts, this might not be an option. But there are a couple of things you can do.

One trick is to zoom in. If a person's hands are important, make sure there is very little else in the shot. This is easy to do, since you can focus your camera before the presentation

In about 95 percent of the cases I see where the video is blurry or unusable, the problem is not the camera but the computer that's driving the camera.

begins. The less background information the camera has to deal with, the less chance that movement will cause it to blur.

The second thing you can do is to tell the software driving the camera that you need faster updating of the image. Normally, this is as easy as moving a slider bar. The trade-off is that, while your camera will be updating more frequently, your image quality will decline. Hypothetically, you would be able to see the sign-language presentation, but trying to read charts would be an exercise in futility. An option you might consider is to use one of the many programs that combine streaming video with presentation software. In one window you can hold a video teleconference, and in the other you can show charts or demonstrate software straight from your computer. The main trick here is to learn to adapt your settings to the manner in which you are using your cam. Try to find a balance between performance and image quality that works.

If you've tried adjusting the quality settings on the camera, reduced the movement as much as possible, and zoomed in to crop out any background data, you should be fine. But if you're still having trouble recording the image you need, it's time to look at possible hardware problems on your system. Namely, it may be time to upgrade at least one component on your system.

In about 95 percent of the cases I see where the video is blurry or unusable, the problem is not the camera but the computer that's driving the camera. Webcams are amazingly simple devices, very much like early box cameras in their design. Other than the light-sensing chip inside the camera and the software, it's just a box with a lens. In other words, it's difficult to make an unusable webcam. Mostly camera-related problems will crop up in the form of poor quality or damaged CCD chips, and it will be quite easy to tell that this is a problem when your blues end up looking like greens, or everything looks washed out or dark even under well-lit situations. As a general rule of thumb, if the problem is in the quality of moving video images off your cam, then it's more than likely that your computer, not your cam, is the cause of the problem.

More often than not, the computer driving the camera suffers from one of two problems: either the CPU is not fast enough, or more likely, the video card is inadequate.

More often than not, the computer driving the camera suffers from one of two problems: either the CPU is not fast enough, or more likely, the video card is inadequate. Check the requirements on your camera's packaging box to make sure your system's CPU meets the recommended requirements.

A dirty little secret throughout the computer industry is to never trust the minimum requirements. Usually the marketing wing of a company will set the minimum requirements for a product very low so that it will be available for sale to more people. The logic often is, "Well, it does work at this requirement level, only not very well." You will see more reputable companies, or ones run by engineers, now put both a minimum requirement and a suggested or recommended requirement on the box. And it's good to be at least slightly above the recommended requirement, too. But if you're right at the minimum requirement line, then there's a good chance the product is performing poorly because of your system.

The other component that you can upgrade is the video card. This is the card that drives all the video functions for the camera and is highly important for any webcam operation.

Thankfully, video cards are both cheap and easy to install. Getting one with 16M or 32M of video memory makes all the difference in the world. And you will see an immediate improvement in your camera's operation, especially if it's a camera that doesn't ship with its own video-capture card, as most do not these days. But either way, things should improve.

Some motherboards have the video adapter built into the board, and these lower-end models sometimes don't give you an AGP slot in which to put an upgrade card.

Even the highest level, best PC video card is relatively inexpensive. The highest I have ever paid for a video card in the past few years is about $120. And installing one is pretty easy as well.

While you should always check the installation instructions first, it is usually a matter of powering your computer down and removing the case. Unplug the power cord. The adapter card that the monitor plugs into will be your old video adapter card. Unscrew any screws holding it in (there shouldn't be more than one) and gently pull it out of the motherboard. Then insert your new card, making sure it fits all the way into the slot. Reattach your monitor and other cables and put your computer's case back on. When you next power your machine, it will probably ask for a driver disk that should be inside your new video card's shipping box.

This process is a bit more troublesome on a PC than on a Mac. You have to see if your card is PCI or AGP on most PCs. This is simply the type of slot your card will fit into. If you're replacing a PCI video card with an AGP card, you will have to put the AGP card into a different slot. Just make sure your computer has an AGP slot and that it is empty. Even if you have documentation for your computer that says you have an AGP video adapter, you should double-check to make sure you have a slot. Some motherboards have the video adapter built into the board, and these lower-end models sometimes don't give you an AGP slot in which to put an upgrade card. In these cases, you're essentially locked out of upgrading your AGP adapter, so it's always a good idea to double-check.

You probably will be in the same boat if you're using a notebook to drive your webcam, as you don't have a lot of choice in video cards. Since you really can't open the notebook and replace the video chip, you will be stuck with whatever setup you have. You can try moving the camera to a notebook with a faster CPU or more memory, but that's about the extent of your upgrade options.

Before you put yourself through all the upgrade hassle, the best thing to do is to move your camera to another computer with a faster CPU and a better video card. See if the performance of the camera increases. If it does, then upgrading your hardware is something you might want to seriously undertake. But normally, the motion problems can be handled by simply telling your subject to sit still and by increasing the quality of the video using the included software.

Filters: How interesting!

Just because you're using a plain, uncluttered background that is flattering to your subject does not mean you can't have a little bit of fun.

Just because you're using a plain, uncluttered background that is flattering to your subject does not mean you can't have a little bit of fun. For the most part, the tips in this section will be more suited for home users looking to add a little spice to their presentations than for serious businesspeople, though you might find a creative way to make use of some of them.

Often people ask me for tips and tricks to making their webcam videos more exciting. A lot of the tips I give are the same I tell people looking for a way to spice up their cheap point-and-shoot camera photos. Of course, the tips are a bit different. For a point-and-shoot camera, I often tell people to try a colored flash for some cool night shots. Webcams don't have flashes (though I've heard some talk in the industry that one might be in development), so obviously some of the tips will be different.

Even though you don't have access to a flash, you can still have some fun with colored filters. Colored filters can be

For a webcam filter, all you really need is a colored piece of plastic. All those fancy filters you find for 35mm cameras will work, but there's no need to pay that much money.

purchased at a theater store, or found in much more mundane places. Feel free to pick some up at camera stores, too, though you will pay a lot more. For a webcam filter, all you really need is a colored piece of plastic. All those fancy filters you find for 35mm cameras will work, but there's no need to pay that much money.

I normally find mine in hobby stores. Since you only need about one square inch of material, you can always buy the smallest bit of material that is available. If you're lucky, you can find blue and red plastic sheets for less than a dollar. When picking out your filters, beware of exceptionally dark plastics. If it's too dark, then your webcam may not be able to film correctly through the filter.

Once you have your filter all picked out, cut a tiny square out of your material. You only need enough to cover your camera's lens, and perhaps a bit more for securing purposes.

What you do is take your square of plastic and put it directly over your webcam's lens. Then take a bit of tape and secure it. You have to make sure that the plastic completely covers the lens. Any bit of light that comes in behind the plastic will throw off your effect.

The best way to secure the filter is to take clear tape and wrap it completely around the plastic that overlaps the camera lens, making sure that none of the tape is put in front of the lens. I have seen people cut circles of colored plastic instead of squares, and depending on how much manual dexterity you possess, you might consider this as an option.

Depending on what exactly you are shooting with your webcam, different colored filters can get the point across. I know one chamber of commerce that uses a blue filter on its outdoor webcam to make a popular waterway look more inviting. This is actually a rather common practice in travel magazines and the like when taking actual photographs. And although it does not work as well with a webcam, you can still use filters to get desired effects and generally make your presentations more interesting.

Red and orange filters are probably the most striking, so they should be used very infrequently. The one time I successfully used a red filter on a webcam was during a New Year's Eve party several years ago. We decided to broadcast the entire party to the Web using a webcam. Since this was still in the early days of webcams, this was actually a unique event that was covered by several Web-based news organizations. To make it even more interesting, we told everyone to arrive at the party in costume. So it was like a giant Halloween party, only on New Year's Eve.

Unfortunately, webcams were barely into their first generation at that point, and the camera we had to work with was not very good. The images were coming out grainy and somewhat uninteresting. Watching people in costume mill around apparently is not that interesting.

Although they do not work as well with webcams as they do with actual photographs, you can still use filters to get desired effects and generally make your presentations more interesting.

So we created a red filter using the plastic cover found in some party game. After attaching it to the camera, everyone started to look really scary. Suddenly the party, for all the Web viewers anyway, took on a mysterious quality. Drinks sparkled with an unnatural reddish glow as people moved in and out of the lens.

The chat room we had created for interaction between the virtual guests and the actual ones was ablaze with comments and questions that quickly replaced the few requests "for the blonde to please come back in front of the camera" that we had seen up to that point. It just goes to show that a slight change can mean all the difference between an ordinary and an extraordinary video.

Some advanced cam software packages even let you apply filters automatically, which are applied to the image during processing. This lets you see the effects of the filter without actually having to tape one to the lens.

Another popular variant on the filter theme is the sepia filter camera shot. You may have to go to an actual camera store to find a sepia filter, because there aren't many real-world applications for this type of colored plastic.

Basically, the sepia filter is a piece of light brown plastic. What makes it so special is that, when applied to a camera, it makes all the images appear to be taken with an old-time camera.

So, with a little bit of costume work, you can instantly produce a classic-looking photo. This is basically what all those amusement parks do when they offer to take people's pictures in Western costumes. The costumes, of course, are an important part of the show, but so are the photographs themselves, which I have seen in both black-and-white and sepia tones.

Personally, I prefer the sepia tones because they look more like an old movie, whereas the black-and-white tones are probably more realistic.

If you prefer, you can also set most cameras to display images in black and white just by using their included software. You don't have to use a filter to do this. On advanced cameras it's as easy as clicking a button on your control panel. But every webcam should have the ability to do this. Even the ones that don't are technically able; they merely don't have the option included in their software settings.

Some advanced cam software packages even let you apply filters automatically, which are applied to the image during processing. This lets you see the effects of the filter without actually having to tape one to the lens. Some camera professionals would call this cheating, but from what I have seen, the virtual filters look as good as the ones you tape on yourself.

Perhaps the ones you tape up might be slightly better in quality, but the difference is so slight nobody will be able to tell one way or the other. Also, the digital filters can do things to your images (inverting colors, color matting, and other digital image prestidigitation) that you simply could not accomplish with your old-school physical filters.

The final thing most people can do with filters is called softening, which could actually come into play in both home use and in a professional-type video. Here again, you will probably have to get your filters in a camera store, since I can't imagine that your local hardware shop would have anything close to what you need.

Soft focus or fog filters are used to slightly blur the subject. The blurring effect is not enough to make the subject appear out of focus, but it does make the subject appear with less harsh lines.

Soft focus or fog filters are used to slightly blur the subject. The blurring effect is not enough to make the subject appear out of focus, but it does make the subject appear with less harsh lines. Aging movie stars often insist on softening the focus on them. I know of one instance where the directors of a movie would smear Vaseline on the lens of their cameras whenever a certain star had a shot, or they would put light cheesecloth over the lens to soften the star's image. Sometimes, you don't want a harsh focus on your subject, and a fog filter could be exactly what you need.

The only thing to watch out for when using any filter is that you also increase the light levels accordingly in order to get a clean image. The darker your filter, the brighter the ambient light will have to be to ensure a crisp photo or video.

Most of all, experiment. So long as you don't actually damage the webcam's lens in any way—I would recommend setting a general policy of not touching it—you will be surprised what you can do with just a few pieces of colored plastic or a quick tweak to a cam software package.

Airbrushing mistakes away

With video, there's not a lot you can do if you end up filming a terrible presentation. There are some shareware tools available that will let you make minor enhancements to your video, such as reducing the amount of blue or brightening up the image. But for the most part, if you don't like the way your video looks, your only real option is to refilm.

If you're merely capturing a photograph or one frame of a video, you have a lot more options available for editing. With the right tools, you can take a terrible image and raise the quality to at least a passable level. To do this, it's important first to understand exactly what makes up a digital image. I touched on this a bit in Chapter 6, but users really need to know the gritty details in order to edit the photographs or individual frames of a video correctly.

A digital image is a series of tiny colored dots called pixels. If you have ever seen Mediterranean-inspired architecture, then you might have also noticed colorful mosaics on the floors of certain buildings. These mosaics are comprised of many colored stones that, when placed together in mass, form a picture. Another example is the Impressionist style of painting called pointillism (a famous example is Georges Seurat's *Sunday Afternoon at La Grande Jatte*), where different colored dots are combined to create larger images.

With the right tools, you can take a terrible image and raise the quality to at least a passable level.

A webcam image is exactly the same thing. It's just thousands of tiny elements all bundled together. Since they are so small, your eye sees them as a complete image. And even the most basic digital image-editing programs will let you fix up images in both simple and detailed ways. You can even use the zoom function found on most digital imaging programs to look at individual pixels of your digital image.

So editing the image, even at such an advanced level, is quite easy. Say you don't like a blemish on the face of your subject. Simply drill down to the actual pixel level of the photograph and remove all the pixels that make up the blemish, replacing them with normal skin color.

One of the most powerful photo tools found in any digital imaging suite goes by many names, but is most often called the Color Sampler. It is often represented by an eyedropper icon.

This is done by using one of the most powerful photo tools found in any digital imaging suite. It goes by many names, but is most often called the Color Sampler. It is often represented by an eyedropper icon. What you do is place the eyedropper over a pixel or area that you want to reproduce. The system then samples the exact color on that particular spot. Once you have the exact color sampled, you can select a brush tool and cover small areas of detail that you don't want to appear in the photograph. It will appear just like the surrounding area you want sampled.

A few years ago, I had a promotional photo taken of myself using a digital camera. The photo I wanted to use was a picture of me sitting in my office. Unfortunately, I hadn't considered examining the background before the photo was taken. It turns out I was in the process of moving some hardware I had recently reviewed out of my office. I had obtained some boxes from a friend of mine who was managing a local toy store. I had piled much of the hardware inside the boxes and stuck them under a table when it was time to take my photo.

When the image came out, it looked good, but several days later I realized that you could clearly read one of the boxes sitting behind me. Because of the angle of the photo, you could not crop out the box and still have me sitting in the photo. The box was not too bad, but you could clearly read "50 Barbie Dolls, Malibu Barbie" on the side—not exactly the image I was going for. If this were a regular photograph, I would have been stuck with the option of either using it, or taking a new photo.

On any photo program, you can increase the light levels on an image with the click of a button.

But remember, a digital image is merely a collection of pixels, just like the colored stones of ancient mosaics. If an ancient mosaic accidentally was created with Malibu Barbie in it, I'm sure the creators would simply remove the offending stones. That's exactly what I did to the photograph.

I opened up a digital image-editing program and sampled the exact color of the box in question. Since cardboard boxes are uniform in color for the most part, this bit of techno-wizardry was surprisingly easy to perform. I simply painted over the pixels with the writing on it, replacing those pixels with the cardboard color instead. In less than a minute, it was just a plan unlabeled box sitting around my office.

Now, I could have done some even more impressive things with the box. Since I had sampled the color of the text before I covered it up entirely, I could have very easily put my own text on the box, making it say something like, "Danger: Thermonuclear device." Once you have the digital photo loaded, you have access to all the pixels that make up the image. And that makes anything possible, with a little practice.

Simpler manipulations are probably going to be more of the norm for most users. On any photo program, you can increase the light levels on an image with the click of a button.

This is accomplished by raising the brightness on every pixel in the image. Often, an image that is just slightly too dark can be enhanced this way, though darker images most likely can't be fixed, since you have to have at least a halfway viewable image for the basic enhancements to work.

I have found that, for the most part, afterimage enhancements tend to work but are much less effective than actually taking the time to set up and compose a good shot. Turning up the light levels on your subject while you film is far better than trying to artificially light the room once the shooting is over.

For video, you don't have nearly as much control. Essentially, the mosaic imagery for a video feed taken with a webcam is the same, except the image is constantly moving. This makes editing one "frame" of the overall production ineffective, since the viewer won't be able to notice the change. It will fly by too quickly.

And to edit the image enough so that viewers can actually tell what you have done requires hours of work using standard programs. About the best you can do is raise and lower the light level, add time or date or signature stamping, and change the speed of the production.

Getting it right before you have to edit is best, even if this means stopping your presentation and starting over—an option not available to you when you're filming a live performance. Mistakes made early on are not easily fixed once completed.

The many concepts discussed in this chapter may seem like a lot to remember for an aspiring webcam-video director. But really, after a few tries, it will become second nature.

Success follows success

Taken as a whole, the many concepts discussed in this chapter may seem like a lot to remember for an aspiring webcam-video director. But really, after a few tries, it will become second nature. Just try to concentrate on mastering one or two elements each time, such as getting the lighting correct and providing a simple background. When you have that down, then try setting up a video that has some type of framing. Move your subject around the imaginary grid to see which line looks best when using the rule of threes.

I'm told that the best Hollywood directors can walk into either an indoor arena or an outdoor setting and instantly tell how many lights will be needed, what about the background will have to be moved or changed, and where exactly they will position their main cameras. Speaking from experience with webcam filming, I can't say I've become quite that good. But I can enter a room and instantly tell if the lighting is right and what type of light should be used for the object or subject being filmed. And I'm getting better at removing distracting objects from the shot. I've still got to work on my rule of threes when it comes to video, especially when I'm the subject. I have a tendency to scorch toward the middle and obliterate my imaginary grid pattern.

At least practicing won't cost you anything but your time. And you can get instant feedback (the video itself) with which to critique yourself. Be a harsh critic, learn from your mistakes, and keep trying. You will find your ascent into the master ranks of webcam-video producers inevitable over time.

Also, this is a great task to get the whole family involved in. You might be surprised by the insights that can come from your spouse, your children, or—and yes, it occasionally happens—even your parents. Create a family-video holiday card, or have your children produce their own short movie. Not only will it be fun, but it can give you tons of practice.

Be a harsh critic, learn from your mistakes, and keep trying. You will find your ascent into the master ranks of webcam-video producers inevitable over time.

Video E-mail: Putting Yourself in a Digital Envelope

Chapter 9

Video e-mail eliminates almost all the drawbacks of typical e-mail communication, but unfortunately comes with its own bag of new problems.

Those of you who have become masters at e-mail probably are acutely aware of both the advantages of this near-instant method of communication as well as some of the drawbacks. Video e-mail eliminates almost all the drawbacks of typical e-mail communication, but unfortunately comes with its own bag of new problems.

We all know the advantages of e-mail. Once you have the connection established, it's basically free. And free beats having to put postage on a letter sent through the regular mail. It's also less of a hassle, since you only need to address the To: field of your mail to send the message anywhere in the world almost instantly.

E-mail, as a means of communication, is often conversational in nature. Unlike trying to compose a letter on paper for someone, you can be freer with your grammar in an e-mail, and most people generally take a much less formal tone. With a paper letter, you basically have to make sure you are 100 percent accurate before you mail it out, since it can be a long time before you hear back from your recipient. But if a person doesn't fully understand what you wrote in an e-mail, he or she can simply hit the Reply button and instantly send you feedback. Even if a person is on the other side of the planet, chances are there is going to be very little time (mere seconds in most cases) between the time a note is sent and the time the receiver gets the mail on his or her computer.

I'm reminded of a coworker who was very new to the world of computers. To her, it was unusual that people at the office sent e-mail back and forth as the primary method of communication. She thought it would be better to simply walk down the hall and talk to people face-to-face.

Unlike trying to compose a letter on paper for someone, you can be freer with your grammar in an e-mail, and most people generally take a much less formal tone.

But another advantage of e-mail is that it's less intrusive than talking with a person directly, so she became quite unpopular. With e-mail, you can simply continue talking on the phone or whatever and get to it when you have a free moment. But when someone actually stops by your office, you pretty much have to deal with that person right away.

Another problem this unfortunate coworker had was that she didn't quite understand how conversational e-mail was as a communications form. So she would take an hour to compose an e-mail, reading over each sentence time and time again to make sure her grammar and punctuation were flawless. And this was done just so could tell her coworkers that she wanted to head to lunch! Eventually, she realized the many advantages of e-mail and became more accustomed to how and when to use it.

But e-mail is far from perfect. If e-mail were a perfect form of communication, then there would be little need for video e-mail, phones, or anything else.

The main disadvantage with standard e-mail is that it's nearly impossible to convey emotions. Thus people have taken to creating little "emotes" (also called "emoticons"), or basic symbols of smiley or sad faces, to represent feelings. But the emotes are restricted to the characters available on a keyboard, so they're very sparse. It's the equivalent of cave drawings in the modern age. Using emote logic, when you get a nasty note and there is a little happy face drawn at the end of it using a colon, dash, and greater than sign (:->), you're supposed to realize that the person was just joking.

The main disadvantage with standard e-mail is that it's nearly impossible to convey emotions.

Obviously, this does not work too well. For one thing, the person on the receiving end has to know what all the symbols mean, especially the more complex ones that supposedly have more advanced meanings. Secondly, many people leave out emotes altogether, assuming that the reader knows when they're being serious and when they're being silly.

I myself have sent a perfectly nice note to a friend (both of us are e-mail savvy), only to have it taken the wrong way. This lead to a bickering match, also known as an e-mail flame war. I got a note back from my friend insulting me, and I wrote back asking what the hell his problem was, which escalated matters. And, like two great countries going to war over a misunderstanding, things got out of hand. With e-mail communication, this is a surprisingly common occurrence. Thankfully, my friend and I made up once we realized what had happened. But the moral of the story is that sarcasm is not a form of humor that can be used effectively in e-mail, so give your friends the benefit of the doubt when chatting using text-only communication. While e-mail may be conversational in nature, remember that there is a lot of unspoken communication that goes on in face-to-face communication that simply can't be replicated in a text-only message, no matter how many happy faces you put in it.

Video e-mail is the answer to many of the problems of standard e-mail. When using video e-mail, you have the option of sending what basically amounts to a one-sided conversation. The person receiving the video e-mail has the advantage of looking at visual cues and examining your body language to help determine meaning. So if you say, "You better stop fooling around and get to work," and you happen to be laughing at the time, the person on the other end won't think you're seriously commenting on their work ethic.

Also, using video e-mail, you can show things beyond just yourself. If you really want people to purchase a certain type of cup as a giveaway item for a trade show, you don't have to describe it to them—simply hold it up and let them take a look at it.

So, video e-mail has many of the advantages of face-to-face conversation, yet is still unobtrusive in a way that's exclusive to e-mail. And it can be done in a variety of ways using standard webcam software. You actually don't need a specialized program to handle video e-mail, though with the variety of attachment formats and the like it is probably a good idea to get some.

Video e-mail has many of the advantages of face-to-face conversation, yet is still unobtrusive in a way that's exclusive to e-mail.

Is video e-mail a perfect communication tool? Nope. Along with all the advantages it brings to the e-mail format, it also has its own unique set of problems that users need to work around.

The primary problem is bandwidth. I've talked about bandwidth in other chapters, and here, too, it rears its ugly head. Your main concern with video e-mail is going to be how much bandwidth the computer making the video has access to, but a courteous person also takes into account how much bandwidth the receiving person has available.

Another major problem is sound, which becomes highly important in video e-mail, since it's often sent without any accompanying text. Poor sound quality is going to garble your message, so you need to take steps to ensure clear audio.

Two other important issues are attachment formats and confirmation of message delivery. These are less important than the first two, but nonetheless something to consider when you're planning your message, especially if you consider the information you're sending to be critical.

Working within bandwidth constraints

A "small pipe" for data can really mess with communications, especially over long distances.

The single greatest enemy to effective connectivity in the modern world is bandwidth. A "small pipe" for data can really mess with communications, especially over long distances. A friend of mine from Tanzania, Africa, came up with a great analogy for the bandwidth problems of the Internet. To improve living conditions, the government of his country built a huge pipeline to carry drinking water across the plains from a reservoir to his village.

Before the pipeline, people had to walk more than 10 miles to get drinking water. And the African continent, especially in the area affectionately known as "the bush," is not the safest place in the world. Sending children through the bush is not always the easiest of tasks, so after years of begging, and a platoon of Peace Corps workers, the government decided to build a pipeline to bring water to the village. Lions and venomous snakes don't bother attacking pipelines, so the

Some people don't consider bandwidth an issue when dealing with video e-mail. But these people are wrong.

village was quite happy at the prospect of getting water without risking life and limb.

My friend said the water pipeline to his village is a like an Internet connection. At first, everything was fine, and the village had a steady, reliable supply of water. But as more and more people living in the plains learned about the pipeline, they began to tap into it, running small personal lines to their homes or larger lines to family complexes.

Slowly but surely, the flow of water at the village end began to decrease, until barely a trickle came out. At nonpeak hours, when people who tapped into the pipeline were not draining as much water, the flow at the village increased somewhat. But in the middle of the day, they were lucky to get a glass of water out of their large pipe.

The Internet is a lot like the Tanzanian water pipeline. As more people get onto the Internet, there is less available bandwidth for everyone to share. This is offset somewhat by the building of new and faster lines, but the larger the pipe you have into the system, the more bandwidth you're potentially going to get out of it.

Some people don't consider bandwidth an issue when dealing with video e-mail. After all, it's not like a videoconference where low bandwidth is going to kill your connection. With video e-mail, you record your presentation first and then simply e-mail it to your receiver.

But these people are wrong. If you don't consider bandwidth issues when planning your video e-mail, not only are you going to make a lot of people angry, but you also are going to end up wasting a lot of personal time and computing cycles.

Consider the following. Say you make a video e-mail that is several minutes long and ends up being close to 100 megabytes in size. On a standard 56K modem, it's going to take several hours for you to send your e-mail. If the Internet is exceptionally busy that day, it may take even longer. Even many corporate connections would have problems sending a 100-megabyte file quickly when everyone else is on-line as well.

Then think about the person who is getting the e-mail. If they have a 56K modem, it will take them several hours to download your message. Depending on the e-mail service they use, they may have no choice but to download your mail before they can get to all the other mail that had been sent to them after your mail arrived.

So you're taking almost a full working day simply to send and deliver a message that is just a few minutes in length. Even though it's not a live performance, bandwidth does matter. Based on the time it takes to send and receive the video e-mail, you could eliminate the primary benefit of e-mail—namely, that it's a near-instant communication. In this example, you are almost back to the letter-writing format for paper, where everything has to be perfect due to the long pause between when a letter is sent and when a reply can be given.

Even though it's not a live performance, bandwidth does matter. Based on the time it takes to send and receive the video e-mail, you could eliminate the primary benefit of e-mail—namely, that it's a near-instant communication.

Other than with a personal or corporate firewall, you can't control the data that flows into your computer. So let's start with the computer that is actually composing the message, since this is the only variable that you will have control over when using video e-mail. How long your message can realistically be will be based on what kind of connection you have to the Internet, or the size of your pipeline. If you have a standard 56K modem, you're going to want to keep your message under 10 MB in size. This can be done by simply keeping your message short and to the point, or by decreasing the resolution that your webcam is recording. If you've made your presentation as short and simple as possible but you're still tipping the scales at over 10 MB, you might try the reduced resolution option. If you've followed all my advice on capturing video in Chapter 8, you should be able to easily use the camera's included software to reduce the resolution, thereby drastically lowering the size of the presentation without a sacrifice in quality.

Another option is to use your video to highlight the points in a presentation that you can either create and send along or post to your web site, and then to send only your message and the link to the presentation. The benefit here is that you still get all your data across and can use your video message to punch up or expound on key concepts.

Even if you are able to send out your 50-Meg message in a few minutes, if the people getting the e-mail have a 56K modem, it's still going to take hours to download.

If you have access to higher connections, such as a cable modem or an ISDN line, you can probably feel free to let your presentation climb as high as 25 MB. The only problem is that few programs tell you exactly how large your video e-mail will be until you're finished. Eventually, if you do it enough, you'll get a feel for it. It will become second nature for you to know ahead of time that, if you talk at a certain resolution for a certain period of time, then your file will be a certain size. Until then, just note how long you talked, what resolution you had the camera set at, and then see how big the file ends up being.

If you have a DSL line, which currently is one of the best home-user connections available, your files could be 50 MB or more. But don't forget about your poor receiver. Even if you are able to send out your 50-Meg message in a few minutes, if the people getting the e-mail have a 56K modem, it's still going to take hours to download. And if it's a frivolous communication, they might be a bit miffed that you wasted so much of their time.

Also, there is the chance that a large e-mail might bounce. Some servers have limits on how big attachments can be, and video e-mail is viewed as an attachment by mail servers. So make sure there are no restrictions where you're sending your mail, or that your mail is under the size limit, otherwise the server won't let it through the gates.

Most webcams don't come equipped with high-quality microphones. Many are totally functional, but you might want to invest in a good headset microphone.

Do you hear what I hear?

The most important component to any video e-mail is the sound. Yes, for once video is not actually the most important component in a webcam application. If you think about it, even if you send a video e-mail with a terrible, blurry video, the message is still going to get through as long as the sound component works. That's because the sound component in a video e-mail takes the place of the text in a standard e-mail. All the video does is give the user clues as to what you mean when you say certain things, and perhaps show the user exactly what you're describing. But the sound carries the information, so you need to make sure that what you're saying is what the user is able to hear.

Unfortunately, most webcams don't come equipped with high-quality microphones. Many are functional, but you might want to invest in a good headset microphone if you plan to do a lot of work with video e-mail. Even if the camera comes with an embedded microphone of excellent quality, you're going to have to set it fairly far away from the person who is speaking, since the microphone is going to be where the camera lens is positioned. If you move the embedded microphone forward, then you have no choice but to move the camera forward as well, which will often throw off your focus—not to mention the composition of the shot.

If you decide that you want to use a secondary microphone instead of the one included with the camera, setup of this secondary device is relatively simple. Just make sure you get a microphone that is specifically designed for use with a computer, because it will have the right connection to plug into the back of a soundcard. Any microphone will work, but those that can also be used with voice-recognition programs often work well. If you buy a microphone for your stereo, for example, you might have to purchase an additional adapter to make it work on your computer. Perhaps not, but I have a box of microphone adapters because I wasn't careful, and you might as well be sure (unless you want to try to build a collection to rival my own).

Many computers come with microphones that mount on your monitor or have their own stand. These will work, but the headset microphones usually provide the best overall sound quality (see Figure 9-1). The only disadvantage is that you have to wear the headset during your presentation,

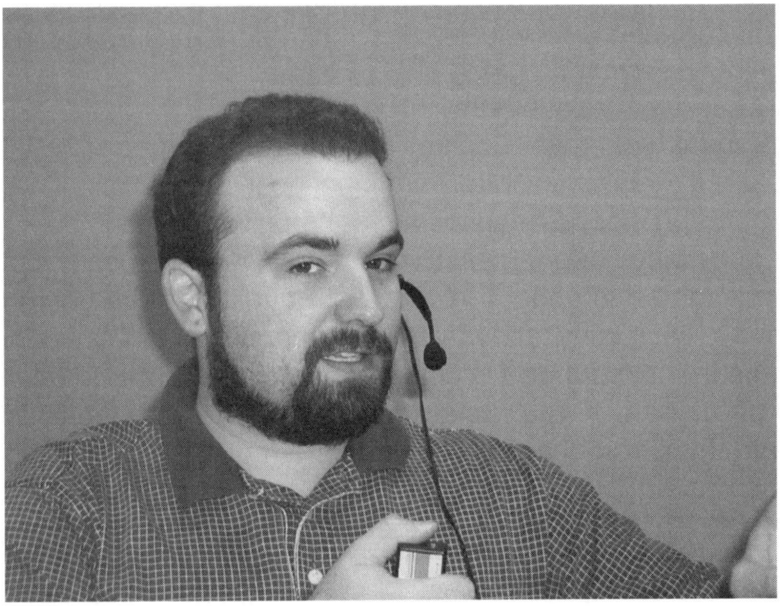

Figure 9-1. An example of a headset microphone, which provides the best overall sound quality.

though I don't really think this is a bad thing. Also, if you plan to use your webcam to do video chat, a headset microphone is almost a necessity. Using a third-party microphone with your webcam is easy. Many have a port for an external microphone, but even if they don't you can still use one. Simply find where the microphone from the camera plugs into the computer and replace the built-in microphone plug with your new microphone.

If you plan to use your webcam to do video chat, a headset microphone is almost a necessity.

Some webcams have a single cord that comes from the camera, but they almost always split near the other end when going into the computer. Most of the time, the wire carrying sound data is clearly marked. If it isn't, a simple experiment can determine which one carries sound just by creating brief test messages. When you don't hear sound anymore, you have pulled the correct wire. Simply plug your new microphone into that hole. But since the camera company wants to make sure you put the wires in the right places, good companies will even go so far as to color code this area, so don't anticipate too much trouble.

Some webcams require that a proprietary option card be installed on the computer, though these are becoming increasingly rare as USB devices gain popularity. If you're stuck dealing with a camera dependent on a proprietary card, you will probably have to connect your microphone into the camera's option card instead of your existing soundcard. If for some reason your camera's card isn't compatible with your microphone, try plugging the microphone into the "Microphone In" jack on your soundcard. Either under your operating system or within the camera's own software, you may have to choose your specific microphone-input source for it to work, but it should be fairly simple. In fact, some applications will automatically default to your soundcard's input once the microphone is unplugged from the camera's option card. I have seen a few cameras set up this way.

Door No. 1, 2, or 3?

Another important decision is what type of microphone to purchase. A lot of this will be determined by your needs.

Another important decision is what type of microphone to purchase. A lot of this will be determined by your needs. If you're setting up a video e-mail station in a corporate office for high-level communications, your needs are generally going to be different than someone who is just interested in sending video e-mail to a long-distance sweetheart. And with portable outdoor applications, which are becoming increasingly popular, your sound needs change drastically.

For most home users, the best choice is going to be your standard computer microphone (see Figure 9-2). These tiny peripherals, mostly painted white, snake from your soundcard and are meant to rest in front of you on your desktop.

The standard computer microphone is hardly high-quality. They sell for a few dollars at any computer store, and even many retail electronics stores. They will work for video e-mail, though you need to set it within a few feet of your mouth. Also, if you move your head while you speak, expect the volume of your recorded voice to fade in and out because of the distance between you and the microphone.

Some of the higher-end cameras come with very good embedded microphones that are actually much better quality than the standard computer microphone, but nonetheless can't be moved, since they're joined with the camera itself. The point is to put the microphone as close to the user as possible. The standard computer microphones almost always win out when doing sound-quality benchmarking simply because they can be moved closer to the source.

Another option I touched on earlier is the headset. Certain headsets have noise-dampening electronics that cut down on the amount of background chatter that is recorded. Here

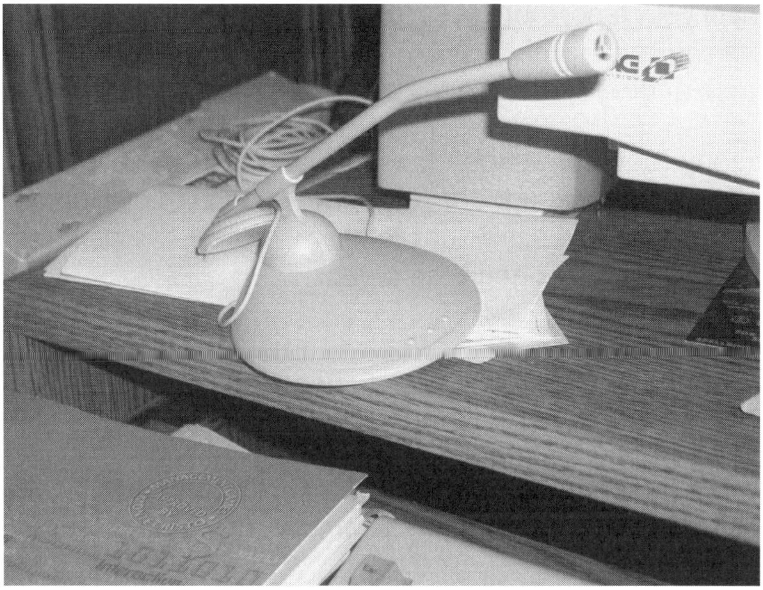

Figure 9-2. A standard computer microphone. Though hardly high-quality, it works well for video e-mail.

you have the best of both worlds: a microphone that sits just inches away from your mouth, and the electronics capability to eliminate noise generated by computer cooling fans, air condition ducts, traffic, or any sound not specifically directed into the microphone. You also don't have to worry about variances in volume, as your mouth remains a set distance from the microphone at all times.

The only disadvantage is that the person has to wear the microphone on his or her head. Some people might think wearing the headset looks silly or makes them look like a World War II radio operator. Also, wearing the headset for a long period of time can be tiring, especially considering that the higher-end headsets tend to be somewhat heavy.

If you do decide to go with a headset microphone, you need to make sure you position the microphone slightly off to one side of your face. The ideal position is an inch or two from your mouth and off to one side. If you put the microphone right in

An increasing number of people are taking their webcams on the road, especially as notebook computers become lighter and batteries last longer. This has led to an expected increase in outdoor uses, including video e-mail.

front of your face, there's a good chance that it will record your breathing sounds, especially in non-dampening models. There is such a thing as getting too close to your source.

A happy (but more expensive) medium is the lavalier microphone. These microphones are popular with television news crews when it is either unwieldy or unsafe to use a large handheld microphone in the field.

Basically, a lavalier microphone is a tiny device that clips to your shirt or tie. It's ideally positioned a few inches below the chin and is designed to record sound coming from right above it, since the microphone points upward toward the person's face. Tiny lavalier microphones are very hard to detect if they are positioned under a shirt, and even the larger ones are mostly unobtrusive. Plus, people are accustomed to seeing them during television interviews and even game shows to supplement the large studio boom microphones that hang down from large poles. Lavalier microphones are basically reverse boom microphones, in that they record up, while booms record what is below.

A fair number of lavalier microphones are wireless, meaning you don't have to worry about getting tangled in a long cord. But the wireless microphones are very expensive to the point of not being practical for most video e-mail applications. Since you don't move much during most video e-mail presentations, you sometimes can save hundreds of dollars getting a microphone with a long wire that feeds back into the computer, instead of a wireless model.

An increasing number of people are taking their webcams on the road, especially as notebook computers become lighter and batteries last longer. This has led to an expected increase in outdoor uses, including video e-mail. Consider people visiting the nation's capital for the first time. They can record a video e-mail on the steps of the Lincoln Memorial for the folks back home and then simply send the e-mail when they plug into the phone line back at their hotel. Really high-tech travelers can send out the e-mail using a wireless network, though the bandwidth is typically very poor when compared to even a standard modem using a landline.

When you're outside, the sound requirements are a little different than when inside a controlled office or living room. Besides an increase in background noise, your primary enemy is going to be the wind.

People don't realize how strong the wind actually is or how prevalent it is when you're trying to use a microphone outdoors. Every television reporter who has ever covered a storm knows that wind blowing across a microphone is loud. And I mean it is loud to the point that you have to really speak up—in some cases scream—just to have your voice projected over the wind noise. I even have seen reporters use this to their advantage, holding the microphone so a lot of wind blows across it.

When you're outside, the sound requirements are a little different than when inside a controlled office or living room.

The end result is that a storm seems to be a lot more dramatic than it actually is. This may seem to lack a certain amount of integrity, but it shows that, just as with the camera, you can use the microphone to add something extra to your digital video recording. But unless you're covering Tropical Storm Humongous, you probably don't want a lot of wind noise in your video e-mail. The best thing to do is to choose a location where the wind isn't blowing. If you can take shelter behind a wall or a grove of trees, you might fare better than standing out in the open. Even putting your back to the wind can help in some cases, like trying to shelter a candle in the wind. But, of course, this is not always possible.

I've found that using a standard handheld microphone is not the most practical thing to do when the wind is blowing somewhat hard. I have gotten around this, however, by creating a shelter for the microphone. The shelter idea came when a friend of mine was trying to create a video e-mail for his wife that was recorded on the lot where their new house was going to be built. The wind was really blowing that day, and an empty lot typically does not have a lot of shelter. So I used a large plastic cup to make one.

Wind noise is only picked up when the wind is blowing across the top of the microphone. So if you cut a hole in the bottom of a paper or plastic cup and slip it over the microphone, it's kind of like cupping your hand around a flame. It prevents the wind from directly hitting the microphone, and you can also more easily hold the microphone in a direction that is out of the wind.

The only other option is to purchase a very expensive specialized microphone, such as a shotgun-type that records at a long distance, and direct the sound recording in your direction from a safe location. And that is more trouble than it's worth, especially considering my free solution works just fine.

You can also use the cup shield with a lavalier microphone, though you then lose the microphone's primary attributes—its small size and portable nature. It's probably better simply to put your back to the wind and hope for the best with one of those.

Wind noise is only picked up when the wind is blowing across the top of the microphone.

Getting attached

Once you have your video recorded, you will have to determine how to get the message from your computer to another person. In the old days, this meant that you had to attach your video to the e-mail, much like you would a photograph. Even today, this is actually what is occurring, but mostly the camera software has evolved to the point that all this is handled in the background and is unseen by the user.

One of the problems from the early days of video e-mail that still persists somewhat is the difficulty of conflicting attachment formats. If your camera records the video e-mail in a format that is not supported by the recipient's computer, then that person won't be able to view or hear your message. This tends to annoy people, since they still have to download a very long message, yet can't do anything with it once they have it safely on their hard drive.

One of the problems from the early days of video e-mail that still persists somewhat is the difficulty of conflicting attachment formats.

This problem mostly has been eliminated by two factors. The first is that two main formats have emerged as the primary way to compress video. The second is that clever companies have created proprietary formats for recording video, yet include everything a user needs to view the video as part of the message in the form of a self-executing application.

The most popular compression format by far is the Motion Picture Expert Group format, or MPEG. MPEG is a group of people that meet under the International Standards Organization (ISO) to generate standards for digital video and audio compression. In particular, they define a compressed bit stream, which implicitly defines a decompressor. MPEG meets roughly four times a year for roughly a week each time. In between meetings, members do a great deal of work, so it doesn't all happen at the meetings. The work is organized and planned at the meetings.

If you really care about video formats, then you should know that Motion Picture Expert Group itself is a nickname. The official name is the ISO IEC JTC1 SC29 WG11, which stands for the International Organization for Standardization International Electro-technical Commission, Joint Technical Committee 1, Subcommittee 29, Work Group 11 on moving

If your webcam uses the MPEG format, then it's almost certain that users on the other end of your transmission will be able to view your attachment.

pictures with audio. But all you really have to know is that 90 percent of all video on the Internet or video used by webcams is put in this format. So if your finished product is saved with an .mpeg extension, then you can be relatively sure that the person on the other end can use a standard reader such as Microsoft Windows Media Player to view your video e-mail.

The other format that has survived the consolidation is Apple QuickTime. Like MPEG, QuickTime describes a format for the creation and distribution of video, audio, and other formats. Although now an ISO standard, it was initially developed by Apple, Inc. for use on Macintosh computers. Today, QuickTime will work on either an Apple or a PC. Although a standard player is free for anyone to download, Apple does periodically update the format. And you normally have to pay about $30 to get the latest full-featured version.

QuickTime has survived for two reasons. First, Mac users are fiercely loyal to their platforms, so it has an advantage there. Second, many people claim that QuickTime movies are far superior to MPEG in both quality and compression. I can't really say which format is better. In my experience, the compression rates are nearly identical. But I have seen better quality movies in QuickTime format, though this could just be random.

In any case, your camera is probably going to save your video in one of those two formats. If it uses the MPEG format, then it's almost certain that users on the other end of your transmission will be able to view your attachment. If it uses QuickTime, however, it might be worth your time just to make sure they have QuickTime capability on their end. If not, they can always download it for free, but this could end up taking a lot of time over a slow modem—and that would counter all the work you did to keep the video as short as possible. Plus, it's always best to make sure the recipient has a soundcard.

In addition, your camera's software could be recording a video e-mail using a proprietary format. Most webcam-software bundles won't do this when recording a normal video, but I have seen an increasing number use their own format when recording a video e-mail. My guess is that the company is doing its best to compress the video, and in most cases considers getting the video down to a manageable size more important than using a standard format.

If your camera does this, however, then you're actually in luck. Since companies know that sending an unreadable e-mail is not popular, they have found a way to include a reader with the e-mail. How this works is that your video e-mail is transformed into an executable file—with an .exe extension—when it's sent to other users. When other users get the e-mail, they only have to click on the attachment to launch the video. And it doesn't matter what kind of computer or operating system they're using, because all the tools to view the e-mail come embedded with the message itself.

The only disadvantage I can think of in using a camera with this type of setup is that the tools themselves could increase the overall size of the e-mail. But in my testing they don't, at least not by a significant amount. Also, many users are understandably nervous about running any file with an .exe extension, as it could possibly carry a computer virus.

Since companies know that sending an unreadable e-mail is not popular, they have found a way to include a reader with the e-mail.

I gathered three cameras that used the proprietary video e-mail method and recorded several e-mails but did not send them. I then measured the size of the raw e-mail and sent them to myself. In every case, the e-mail size was only increased by at most 50K, which translates to a few seconds even on a slow modem.

So even though this format would seem to be a bit of a negative on the surface, it actually clears up a lot of issues when

The Internet is designed to be redundant, so there's almost no chance that an e-mail sent out will disappear without a trace.

dealing with attachments. Just be wary of any camera that saves files in its own proprietary format and then doesn't include viewing tools when sending out the message. That is a recipe for trouble that most companies have dropped, though it was common at one time.

Over and out

So you've worked for hours getting your video e-mail just right, got the format worked out with the person you were sending it to, and pushed the Send button on your console. Now what? A lot of users actually panic at this point, and I've seen it firsthand. They wonder if the video e-mail actually got through to its intended destination. They fret that it will get lost or broken somehow as it travels through the somewhat magical highways and byways of the Internet.

Actually, this is the simplest part of the equation. The Internet is designed to be redundant, so there's almost no chance that an e-mail sent out will disappear without a trace. And this includes a video e-mail, which is basically just a normal e-mail with a video file attachment.

In its simplest form, the Internet can be described or diagramed as an old-fashioned fire brigade going between two computers. Your file is the bucket of water being handed down the line of firefighters. The difference is that the Internet is actually millions of fire brigades crisscrossing the world, and your bucket only ends up in the right place because it has its destination address written on the side. What happens is that your e-mail is encoded with data at the very top of the file that tells routers—the equivalent of Internet traffic cops—where your final destination is located and where the mail is coming from.

When you hit Send, your e-mail goes to your Internet Service Provider's mail server. This computer basically decides whether to send the message internally (to another user of your ISP) or on to the mail server of the addressee's ISP. If it goes to another ISP, it looks up the address of the ISP and sends the message to a router (think of this as the first firefighter in the brigade), normally the last one controlled by your ISP. From then on, your message is out on the Internet.

The message is broken down into small chunks of data, each with the intended address encoded within. The router then picks another router to send your mail to down the line. The next router, which is in the general direction your message wants to travel, then decides to keep it or pass it on. One of the truly astounding things about the Internet is that those chunks of data that your message was broken up into may travel different routes before ending up at the destination mail server, where they are reassembled in the correct order. Think about this happening billions of times a second, and you begin to understand the full weight of what the Internet means.

One of the truly astounding things about the Internet is that those chunks of data that your message was broken up into may travel different routes before ending up at the destination mail server, where they are reassembled in the correct order.

Sometimes a router might be busy with another application, such as a videoconference that has been given a higher priority by the system administrators. If this is the case, then your message might be delayed somewhat.

All this sounds convoluted, but in terms of the real time it takes for e-mail to reach its destination, it's basically instant. A router or a server along the chain might have to delay your message for a few seconds here and there, but your mail is still going to arrive in under a minute in most cases.

I actually did a cross-country experiment once to prove this theory. I simply called a friend on the phone who lived on the other side of the country and had him boot his computer and connect to the Internet. Then I sent mail of varying sizes and timed how long it took to reach him, not including the uploading and downloading times for very large files. I found that the most time it ever took to deliver a message across the nation was two-and-a-half minutes. And the average time it took was just under 10 seconds.

But the great thing about standard mail protocols is that they're redundant and keep the sender informed about major bumps in the road. If for some reason your message gets stuck—normally on your recipient's mail server—you will get a warning message after a period of time (about four hours). Whenever the problem is fixed or fixes itself, the mail will just go through normally. After one to three days of repeated delivery failures, your mail will come back to you with the general reason that it could not go through.

Also, if you send the mail to a bad address (say you mistype your friend's e-mail), it will almost instantly bounce back to you with the note that no such user exists on the server that was the intended destination. No matter what happens, your mail will either be delivered or you will know that it didn't make it. So whenever someone claims not to have gotten an e-mail from me, and I have a record of it being sent, I'm always a bit suspicious. E-mail generally never gets lost without a trace. The very structure of the Internet prevents this.

No matter what happens, your mail will either be delivered or you will know that it didn't make it.

The best way to make sure your mail will get to where it is going is to double- and triple-check the address of the person who's getting the mail. Especially with video e-mail, which can take a long time to upload, a tiny bit of preventative maintenance can save you the time and hassle of figuring out where you went wrong and forcing you to resend the document.

So relax. As long as you typed the right e-mail address into your mail program, your video e-mail will reach its destination.

Internet Video Chat: Videophones

Chapter 10

The idea of a videophone has sort of jumped to a different track.

Would somebody please answer the computer?

For decades, prognosticators have been forecasting the widespread use of videophones. As originally envisioned, these would take the place of common telephones and provide a way to transmit and receive both voice and data. In the usual fashions of such predictions, truth is stranger than fiction. While there are limited units out there that provide still images and voice, and even some that provide video and voice, the idea of a videophone has sort of jumped to a different track.

With an Internet connection, a webcam, and a microphone, you cannot only place video "calls," you can have on-line whiteboard diagrams, share files, and even watch as software is demonstrated. At the moment, this places computer videoconferencing a bit ahead of what was envisioned, and also a bit behind. Video and sound quality aren't nearly up to the TV quality that was prophesized, especially over a 56K dial-up connection. But with faster connections, better imaging- and audio-compression algorithms, and of course, faster computers, audio and video streams will actually exceed that of TV signals in the not-too-distant future.

The exciting thing about this concept is that, by taking advantage of your investments in your computer, your Internet connection, and your webcam, you can get in on the ground floor, as it were. And all those video calls to friends and family members are free, outside of the investments you've already made.

If you are among the millions of people who either already are or are considering telecommuting, a webcam and a speedy connection can bridge the gap in communication between you and your colleagues.

Not only that, but if you are among the millions of people who either already are or are considering telecommuting, a webcam and a speedy connection can bridge the gap in communication between you and your colleagues back at the office or anywhere around the world.

Rather than sitting around a conference table, sipping bad coffee, and wondering who called the meeting, you can take part in a quick and painless videoconference, engage in the exchange of information that you needed, and get on with your day (see Figure 10-1). For the moment at least, people participating in videoconferencing tend to stay more on topic than when in actual conferences. There's the realization that the meeting is using up resources, taking bandwidth and your time.

As the technology matures and becomes more commonplace, you can probably count on this attitude dying out somewhat. Something very similar happens with every new advance in communication. When the telephone was first introduced to

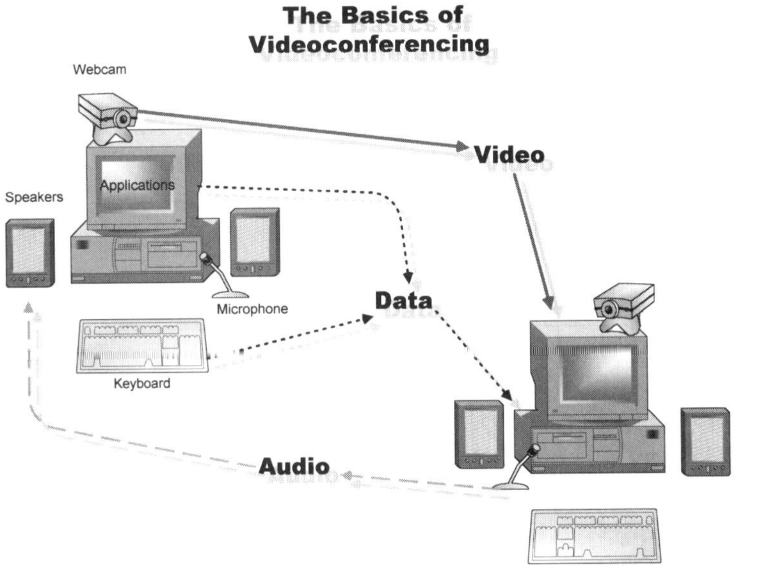

Figure 10-1. Diagram of computer-based videoconferencing, which has at its heart the real-time transfer of video between users. It also features sound and text streams, application sharing, and virtual whiteboards.

America, people saw it as a device only to be used for special occasions and emergencies. Obviously, that is no longer the case, and the change happened very quickly.

For home users, the use of the computer for chatting by voice and video over the Internet can bring a great many benefits. Keeping in better communication with faraway friends and distant family is only one. For people with disabilities, such as the blind, who have been largely left behind by the rising tide of Internet communication, it can mean a new way to communicate to the world. Also, for people living in remote areas, it can be an inexpensive way to maintain a link to loved ones.

As you can tell, this is one of the applications of webcams that we find very exciting. For it is this use of webcams, more than any of the others combined, that will potentially bring a digital video camera to every computer. It is what is referred to as the "killer-app." This term is used to describe applications that have shifted—if you will excuse the use of a cliché—paradigms. For the early personal computers, it was spreadsheet applications such as VisiCalc and word-processing programs such as WordPerfect and WordStar. More recently, we've seen the Internet cause an explosion in the purchases of modems and other means to connect to the World Wide Web.

The ability to talk to *and* see anybody around the world for a price equivalent to the amount of your daily pocket change will be enough to drive a huge upswing in the use of webcams. So why hasn't it happened already? Well, obviously it has a little bit, as evidenced by the fact that you bought this book and are reading this chapter. But the big "bump" isn't going to happen until the final linchpin is in place, and that is high-speed bandwidth to the Internet and an Internet infrastructure to support faster connections.

Currently, holding a videoconference over the Internet is a shotgun-like exercise.

Currently, holding a videoconference over the Internet is a shotgun-like exercise. Your computer, microphone, and camera push out as much information as possible, and on the other end you hope some of it hits the target. This results in choppy-looking video and audio streams interrupted by distracting pauses. And as videoconferencing and other such bandwidth-hungry applications grow in popularity, the load is going to tax a system and an infrastructure that will be hard-pressed to keep up. Already much of what is considered the "backbone" of the Internet is growing obsolete or overloaded.

In fact, the scientific and research groups that took a fledgling network of computers and hardware and helped it grow into the Internet we know today have already begun to create a new, faster network to allow them to exchange the massive amounts of data necessary to their work. But the Internet everyone uses is also being updated, though in a decidedly more piecemeal fashion. So while videoconferencing over the

Internet is sure to be the "thing" in the near future, it is still a growing, changing modality. In the meantime, however, you can keep in touch with loved ones, have a lot of fun with it, and maybe even get out of that long drive to the office! Just as long as you keep in mind the technology's current limitations, you won't be disappointed. So here are some tips and tricks and all you need to know to use it well.

How fast is fast enough?

You know those signs on some highways posting a maximum speed *and* a minimum speed? If you plan to use videoconferencing, think of the famed "information superhighway" as having those signs. A 56K dial-up is the bare minimum you should consider, and if you use it often, that 56K will seem about as fast as a snail's pace.

While you certainly can use it with a 56K modem, I really recommend that if you plan on taking advantage of this technology, you should consider a connection that provides at least 128K upstream and downstream speeds. That means ISDN, cable modem, or some flavor of DSL.

If you're traveling on the computing equivalent of the autobahn, make sure you have enough power to keep from getting run off the road.

Horsepower under the hood

In keeping with my highway-driving theme, I should also mention that if you're traveling on the computing equivalent of the autobahn, make sure you have enough power to keep from getting run off the road.

In other words, you shouldn't expect videoconferencing to work well if you attempt it with a bargain-basement, $299 computer.

If you plan to use it as part of a telecommuting setup, your computer will be focusing on too many things to handle video and audio streaming very well, if at all. You don't need to use top-of-the-line equipment either.

On the PC side, I recommend at least a Pentium II 266 MHz (or equivalent) based computer with 64MB RAM. For Mac aficionados, any of the newer Macs will work fine, especially some of the new iMacs that have video-editing software included. For computers using Linux or another Unix operating system, I can't offer you any exact specifications, as all the platforms are different. If you run into problems with any of these platforms, however, my first recommendation is to add more memory and make sure you have plenty of hard-disk space free. Personally, I always try to have at least a gigabyte free. That way, you're sure to avoid any storage problems, and with drives today providing anywhere from 30 to 50GB to start, that shouldn't be too difficult.

Especially in a business environment, you want to come off as smooth and professional.

'I'm ready for my close-up, Mr. DeMille'

Before you get on-line and start talking to people, let's make sure of a few things. All of the following advice is stuff I've already talked about (especially in Chapter 8), but allow me to reiterate it. Setting up the position of your webcam, the lighting in the room, and framing yourself within the view of the camera (see Figure 10-2) are all important anytime you're using a webcam, but doubly so when using it to talk to someone in a chat room or an on-line conference. Think of it as the difference between previously taped shows and live programs on TV. Especially in a business environment, you want to come off as smooth and professional.

So do a couple of test runs by using your webcam to record a video of yourself to a file. With most video software, you can choose different levels of video quality, and for our purposes here, you should set that quality to something fairly low.

Remember that when you're talking to someone over the Internet with your webcam, the quality of the sound and video will be dependent on whoever has the slowest connection in your conversation. Setting the quality of your recorded video to low in this case will allow you to evaluate your setup under the conditions you'll most likely be seeing when you do it for real. If you're using a high-speed connection, and the people you will most likely be talking to have high-speed connections, then you can set the video quality higher.

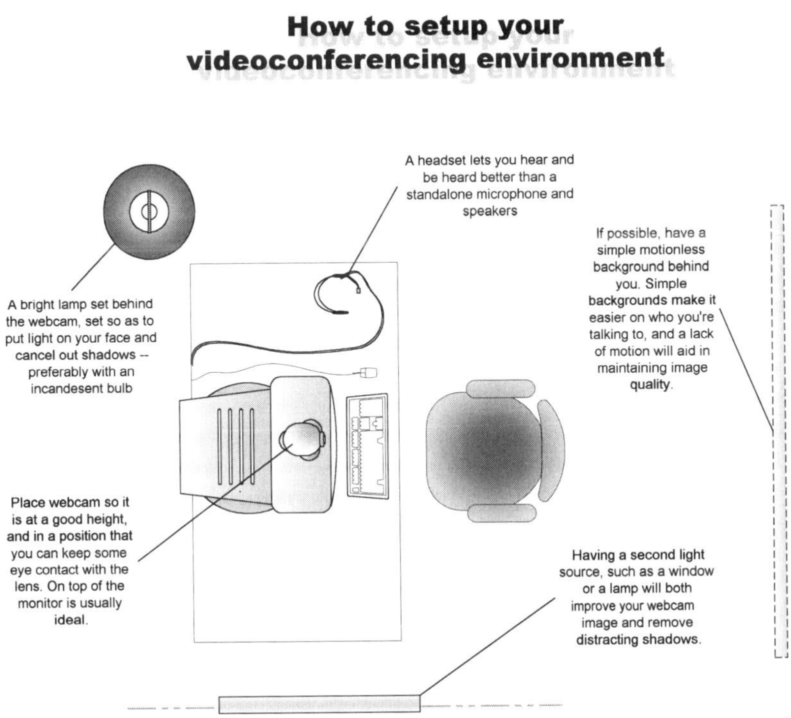

Figure 10-2. Diagram of proper set up for videoconferencing, which can help you make a good impression. Make sure you have adequate lighting and a simple, motionless background. Also, keep background noise to a minimum, since even headset microphones can pick up a wide range of sounds other than your voice.

When playing it back, examine how it looks in terms of video quality and composition, and also pay attention to sound quality. Remember that when you move around too much on camera, your image will become blurred. In fact, even having moving objects in the background can cause image-quality problems. Of course, you also need to make sure your camera is in position so that you stay in the frame of the picture.

When talking to someone in a one-on-one conversation, feel free to have the camera zoom in a little more on your face, but during the often business-related on-line videoconferences, zoom back a little.

When talking to someone in a one-on-one conversation, feel free to have the camera zoom in a little more on your face, but during the often business-related on-line videoconferences, zoom back a little. This is especially true if you plan on being able to show anything other than yourself, such as products, charts, and similar props.

Also examine the recorded demo with your ear as much as your eyes. If you have a stationary microphone located either on your desk or mounted within your webcam, and if you tend to move toward and away from the microphone while you're speaking, you will end up with uneven sound quality. Make sure there isn't a lot of background noise either. You don't want the person to whom you're talking to hear paper shuffling and background noise rather than your end of the conversation. In general, the best solution is to have a headset speaker and microphone setup. This way, you can be assured that you and the person to whom you're speaking are both enjoying the best sound quality possible.

After you've made sure you can be seen and heard as well as possible, it's now time to take the plunge and actually talk to somebody. I mentioned before that there are two different milieus in which to hold video conversations on-line. One is to talk to friends and family, and the other is to talk to colleagues and clients for business. First we'll discuss using the webcam for personal video chat.

One ringy-dingy . . .

A point-to-point connection is just what it sounds like—a connection over the Internet between you and one other person.

Using most commercial videoconferencing software is much like using a phone. To start chatting with people, you first have to know their "number." Depending on the software you use, this might be a user ID or an IP address. Usually, with many of the on-line directories, a user ID is used because it protects your privacy and ensures better security by hiding your IP address. In addition, your IP address will often change if your ISP or office network assigns it dynamically.

Most video software will also allow you to either make a point-to-point connection or connect to a video chat or conference servers. A point-to-point connection is just what it sounds like—a connection over the Internet between you and one other person. This is similar to how a person-to-person telephone call is made. Videoconference servers, on the other hand, are set up as a central connection (called a "reflector") between many different video-chat participants. This server may be a public server, a server within your organization's own network, or a personal server. Anyone who has dialed into a conference call using a regular phone should recognize the concept.

You might be thinking that, when I'm talking about servers, I'm talking about separate computers such as a LAN server. While large public-conference servers may be run off of separate servers, a private-conference server can be run as an application on your home desktop. To do so requires a certain amount of computing power and lots of bandwidth—if you're planning on having more than a few people connected at once—but it's not that hard to set up.

So in summary, to talk to any one person, you'll most likely find that listing in an on-line directory and connect directly to that computer, while if you engage in an on-line conference with three or more people, you'll find a listing of the conference server and connect to that. Once connected, you'll be able to see the other people in the conference. While you can have a greater number of people in the conference, you usually can only see at most 12 people on your screen at one time. This is partially due to bandwidth constraints, but mostly it's just that you can't display that many people recognizably on your monitor simultaneously.

Now that you understand the concepts, what about the details? Well, obviously the methods are going to differ slightly from software to software, but there are some basics that remain the same.

The first thing to do is to go through the steps mentioned above to check your audio and video setup. A lot of the packages on the market have software wizards that will help you through this part.

The next thing to do is to find an on-line directory, or User Location Server as they're called. These can list either individuals looking to chat, on-line conferences, or chat rooms. You can guess that there is not one directory listing everybody, as the logistics would be just too much to handle, and many of the directory servers are of a commercial nature and owned by competing companies. Picking the right server can depend on everything from the responsiveness of the system (many of the directories are often slow and overloaded) to which one has the people you want to chat with listed. To add more confusion to the mix, you may see these servers called by other names, such as Internet Locator Server (Microsoft), Dynamic Lookup Server (Netscape), and other names.

Picking the right server can depend on everything from the responsiveness of the system to which one has the people you want to chat with listed.

No matter what you call them, though, they work by recording your user profile and current IP address when starting up your conferencing software. This makes them sort of an up-to-the-minute white pages. Once you find the person you wish to chat with, or the conference server you want to join, you can simply click on that name.

If you're establishing a personal video chat, most programs will then prompt the person you've selected to speak to and ask that person if he or she would like to talk to you. Once they've selected to speak to you, a point-to-point connection is established and you are all set to talk!

With videoconferences, it can be a little bit more complex. The initial steps are the same, but oftentimes videoconferences have passwords to ensure that only those people who are supposed to be there are allowed to participate. This is a good idea whenever you're setting up a business-related, on-line meeting, as you only want you and your business associates to be allowed in the virtual "conference room."

Oftentimes videoconferences have passwords to ensure that only those people who are supposed to be there are allowed to participate.

Once it's established that you are supposed to be there, a connection is established between you and the conference reflector. This reflector rebroadcasts the feeds from the other participants to you and takes your video and audio feeds and rebroadcasts them as well. Reflectors can be set up to help mediate the feeds, ensuring that everyone is getting roughly the same information in the most efficient manner. In addition, if you're using software that allows you to share on-screen information such as digital whiteboards or desktop sharing, the reflector can direct who has control of those features.

The nice thing about many of these software packages is that even those from different vendors can often work together, thanks to independent standards. The standard for videoconferencing and chat is called H.323, and the standard for real-time sharing of applications and data through conferencing is called T.120 (see Figure 10-3). If your software

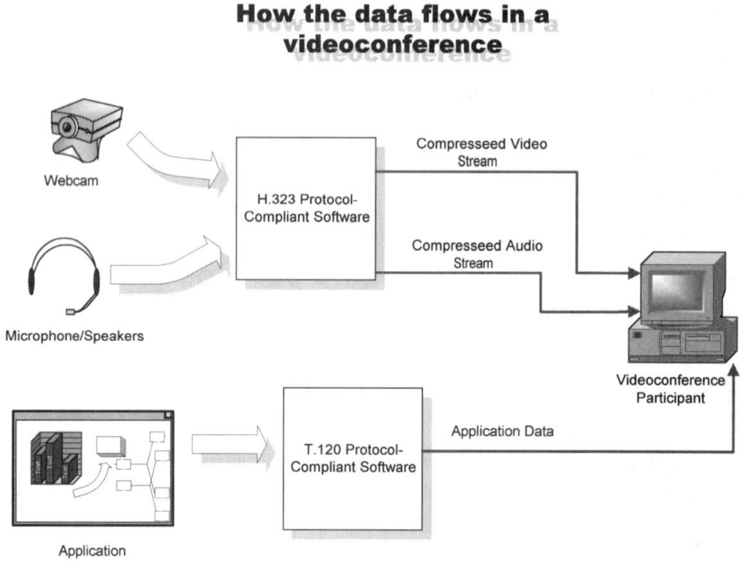

Figure 10-3. Diagram of different data types. In videoconferencing, audio and video data is handled under the H.323 protocol, while the sharing of applications, virtual whiteboards, and other data sharing is handled under the T.120 protocol.

says it supports these standards, then you should be able to use it to conference with any user whose software supports those same standards. Of course, many vendors add features or enhancements not addressed in those standards for conferencing between users of their own products, but that simply means those features are not available when conferencing with users of other products.

One interesting development is the inclusion of video chat in instant-messaging software. Late in 1999, Microsoft stopped having its own ILS servers, leaving that to third-party organizations, and focused on having its MSN Messenger software be the "matchmaker," as it were, for those who wished to video chat. It works in much the same way, recording user profiles and IP addresses when you connect, allowing people to find one another.

Another solution, suitable for any conferencing software, is to use an instant-messaging application called ICQ from

Mirablis. Using this software, you can tell when a friend is on-line, provided you know the user ID and IP address. Using the instant-messaging feature, you can contact them and find out if they wish to video chat. From there, it's simply a matter of one of you initiating a connection from within your videoconferencing software by using the IP address provided through the ICQ user profile.

It should be noted that ICQ does allow users to keep their IP address from being listed (for security reasons). If you're concerned about security, then you could set up your videoconference and then disconnect from ICQ, which should minimize your risk.

Other instant-messaging applications, such as AOL Instant Messenger and Yahoo! Messenger, don't allow you to see other users' IP addresses, so they aren't suitable for initiating video chats. There are, however, even more approaches you can take, if you wish. One is to have a static IP address with which other users can always find you on-line, and use any one of the many instant-messaging programs to determine on-line status.

Some instant-messaging applications don't allow you to see other users' IP addresses, so they aren't suitable for initiating video chats. There are, however, even more approaches you can take.

There are also a number of web sites designed to act as directories for video chatting. One such site is www.meetspace.net, which is a free service.

Another approach is to register with one of the services that assign a name to your IP address. Normally, this could only be done if you had your own domain, but this service gets around that by providing you with a DNS entry and updating your IP address every time you connect to the Internet.

I know of two web sites that do this: one is called www.tzo.com and the other is www.dynip.com. Neither of these services is free, but they are cheaper than registering your own domain.

Also, registering your own domain requires a static IP address, which most home users don't have anyway.

Internet videoconferencing is probably the key to bringing about wide-spread use of webcams.

Looking ahead

Is the day coming when you can throw your telephone in the garbage and connect to the world through your computer, a webcam, and a high-speed connection? Probably not, but that day is in the future. In the meantime, you can use your webcam to communicate with friends and family, as well as business associates.

I've said it before, and I'll say it again: Internet videoconferencing is probably the key to bringing about widespread use of webcams. Even now, though, if you and your video-chat correspondents have high-speed connections, then you can enjoy nearly seamless video and audio streams when conferencing. If you are one of the majority that have a 56K dial-up connection, you can still use video chatting, but quality will be spotty—and don't count on being able to converse with more than one person at a time with any reliability.

One other factor in the coming growth of videoconferencing is the advent of multiuse digital cameras. These cameras are designed to take and store digital still images away from the computer, as well as to act like a webcam when attached. Most digital still cameras today can be used in this fashion, but since they aren't designed to sit easily on a monitor (the spot most people place their webcams) and are sometimes cumbersome to connect and disconnect from the computer, this application hasn't really been taken advantage of. But now there are cameras out that are designed to be multi-purpose. For example, Creative Labs has come out with a camera called VideoBlaster WebCam Go. This camera can be used as a webcam or as a stand-alone digital still camera. Its specifications put it at the lower end of the digital still camera market, as it is only capable of 640 x 480 resolution images, but it's an interesting product nonetheless and a sign of things to come.

So remember my advice about how important bandwidth is to videoconferencing, sign on and find a few friends or make new ones, and chat away!

One other factor in the coming growth of videoconferencing is the advent of multiuse digital cameras.

Cams on the Web: Living the Cam Lifestyle

Chapter 11

So you want to live in a fishbowl?

Perhaps no application for webcams has stirred up as much interest or controversy as using them within web sites.

Perhaps no application for webcams has stirred up as much interest or controversy as using them within web sites. Predating even the famous JenniCam, webcams have been used to provide images for web sites for years. In the early days, it was often an analog video camera hooked to a video-capture card, and then along came digital video cameras.

They displayed images of exotic locales, pets, hectic city-street scenes, and other impersonal images. With JenniCam and others arriving on the scene, however, the use of webcams was taken to a new level. Often using simple JavaScript and a webcam, people were able to design web

Debates raged about what to make of this new use of technology. To this day, nobody—including the people who have webcam sites—seems to have settled on an answer.

sites that presented constantly updated images of their lives to the outside world. At the office or at home, people could watch others in a soundless sort of soap opera, filled with the everyday activities we all engage in: sleeping, eating, working, and sometimes even those more intimate moments.

Some were aghast that people could open their lives to such an extent, while others were enthralled by the idea. Debates raged about what to make of this new use of technology. Was it art, was it pornography, or was it just life? To this day, nobody—including many of the people who have webcam sites—seems to have settled on an answer.

It has been one of those surprises that sometimes come along with new technologies. In hindsight, it's easy to see the trends that led up to it, but it was something even the makers of webcams had not really foreseen. And its use in this way has caused an explosion in interest in webcams.

There are still many web sites devoted to providing pictures of vacation spots, fish tanks, and traffic jams, but most of the traffic to webcam sites now goes to sites featuring people. Interestingly enough, the adult-entertainment industry and amateurs seem to have focused in on webcams as a way to make money. Type the words "adult" and "webcam" into a web search engine and you'll probably get anywhere between ten thousand and fifteen thousand web page hits. This has led to a bias in many people's minds when they hear the word "webcam." But that is an unfair reputation. Go back to a search engine and just type in the word "webcam" and you will get anywhere from tens of thousands to hundreds of thousands of web page hits. Obviously, the majority of webcam sites out there aren't adult in nature.

Some people simply do it to express themselves, others do it as a means to keep in touch with friends, and many instant-messaging users do it as a way to have the people they're chatting with see them without the high-speed bandwidth requirements of streaming videoconferencing. There are also many who see it as a sort of ultimate performance-art piece; simply saying having their lives broadcast over the Internet is the ultimate self-expression.

Then there are also businesses that use webcams on their sites, usually showing something interesting about their business, provide a service, or something else to drive traffic to their web pages. A perfect example is that some day-care centers now have web sites using images from a webcam. Throughout the day, parents can check up to see their children and see how they're doing, giving the parents an added feeling of safety.

In a similar vein, I've heard of kennels providing images of pets to absent owners. I've even seen some stand-alone cameras that automatically update images to a web page marketed as home-security devices.

Whatever use you have in mind for a cam to provide images for a web site, there are certain basic steps that are the same no matter what your goal is.

Everybody say 'Cheese!'

Whatever use you have in mind for a cam to provide images for a web site, there are certain basic steps that are the same no matter what your goal is. First, you have to have your webcam take a picture, and then that picture has to be transmitted to the server (if the camera is not directly connected to a web server) from which your web site is hosted. The page that displays the image has to be written in such a way as to allow the image to be updated whenever a new image is transmitted. It may all sound a little confusing, but it is actually quite simple.

The image capture and transfer is usually handled by software designed for that purpose. An excellent example is called WebCam32 from the Surveyor Corporation (available at surveyorcorp.com/webcam32/). This software is shareware and costs only $25, yet has all the features you could need. It also supports a number of different cam web site configurations.

Another outstanding shareware program that sells for $25 is ChillCam (found at www.thegeek.com/chillcam/), which offers many of the same features and also includes the ability to add interesting photographic-effect filters to your captured webcam images.

The basics of what all this type of software does is to schedule image captures and transfers. The next step in the process is having your web page take advantage of these images.

The basics of what all this type of software does is to schedule image captures and transfers. The next step in the process is having your web page take advantage of these images. For this, it will help to have some knowledge of HTML programming and JavaScript. Scared? Well, there's no reason to be. Thanks to the pioneering work of others, templates for what you'll need to do are easily available on-line.

One particular site I found to be very helpful is called GeneriCam and can be found by pointing your browser to www.occonnect.com/johnqpublic/index.html. This site provides templates for a number of different ways to have your web page automatically update your images, which usually requires you only to edit a little information and copy it into your page.

Another good site to check is www.homecams.com, which provides all sorts of information on setting up a webcam-enabled site.

Now, for all of this to do any good, of course, you have to have a web site to post your pages to.

Stake your claim on the new frontier

The first thing to consider when thinking about starting your own web site is: what are you going to do with it?

What do you mean you don't have a web site? Just kidding! After all, most people today don't have web sites. The first thing to consider when thinking about starting your own web site is: what are you going to do with it? Is it going to be for the edification and entertainment of friends and family, simply for your own enjoyment, or do you plan on letting the world beat a path to your virtual door? All of these can fall under the heading of a "personal" web site. If you're planning on using your webcam to make some money on the Web, check out Chapter 12, which details how to set up a business web site.

Personal web sites can entail differing amounts of resources, energy, and time. A small project might be similar to building a model ship, while other personal web sites are akin to making your own 30-foot yacht. How much time and effort you wish to expend should play a large part in deciding how you want to go about setting up your site.

The simplest way, and often the cheapest, is to use the web-page space allotted you from your ISP. Did you know you have web-page space from your ISP? Most Internet Service Providers provide some kind of web-hosting space to customers in amounts from 3MB to 10MB. It is usually provided as part of your service and costs you nothing extra. You may have to contact them to get it set up, and if not it is still a good idea to check their web page to see how to use it. Often you will be given a URL such as www.yourispname.com/~youruserid. This is the address people would have to type in their browsers to get to your page.

To add or modify pages to your site, you will more than likely have to use FTP (or the File Transfer Protocol) to send them to an address such as FTP://ftp.yourispname/'youruserid.

Your ISP should provide the software that allows you to transfer files via FTP, or you can use a third-party application or sometimes even the built-in tools of your operating system.

Another option is to sign up for free web-page space from one of the many companies offering such services. Their services are similar to those provided by your ISP, in that you get a fairly limited amount of space and there are limits as to how much traffic can be generated by your web site. The interesting thing is that a modest personal home page with all the usual stuff such as information about you, your favorite links, and an updating webcam page can easily fit in 3MB of web-page space—and certainly in 5 or 10MB.

But let's say you want more, such as galleries of old images, software downloads, MP3 samples of your favorite music, and all sorts of multimedia—or maybe you're just planning on having thousands of people come to your site. After all, there's no reason you can't be the next Jenni, is there? This is where it starts getting expensive.

But let's say you want more, such as galleries of old images, software downloads, MP3 samples of your favorite music, and all sorts of multimedia. This is where it starts getting expensive.

To register your own domain name, you'll need around $35 a year. But that only gets you a name; it does not give you a web site. To have an ISP or other company host your site will run you about $30 a month for a limited space and bandwidth (around 30MB of space and 2,000MB a month in data-transfer bandwidth), and approximately $50 for a one-time setup. If you want more space and bandwidth, you're looking at more money—possibly up to $100 a month for about 75MB of space and 6,000MB of bandwidth. That's a bit much, but no matter what levels of server hosting you're looking at, you may be able to help defray the costs with advertising.

Whoever you choose to host your server, you should check out to see what hours they offer support, if they offer you statistics on visitors, and what software they use to host and whether it will work with your web-design software.

While I recommend that you learn as much HTML coding as you can, it is surprisingly simple these days to design web pages. Programs such as Microsoft's FrontPage and HoTMetaL PRO from SoftQuad Software are two excellent packages for this. Using them and other programs of their kind, you can either code the raw HTML files or you can use friendly, easy to use drag and drop tools to design your pages.

The only disadvantage to the drag-and-drop tools is that they sometimes do not create HTML code that's very efficient. For a small personal web site, however, this is of little consequence, yet it's still a good idea to learn at least some HTML.

Giving you instructions on how to code a web site would take a whole other book, but there are many great resources available both in print and on the Web.

If you have any experience with desktop-publishing tools, then programs such as MS FrontPage will seem very similar to you. After all, designing a printed page and a web page are not very different concepts. With both, you have to take into account the size of your page, whether those you intend to read it will understand, and general aesthetics. It gets a little bit more difficult with web pages because of the differing requirements of browsers, operating systems, and web servers. Also, a web page should be more dynamic and engaging then a printed page, which is one of the reasons why adding a webcam image can add a lot to your site.

Giving you instructions on how to code a web site would take a whole other book, but there are many great resources available both in print and on the Web. One of the easiest is to see how the professionals do it. With Internet Explorer and Netscape Navigator, there is a command off of the drop-down menus that allows you to see the source code for any

Once you have your page designed and posted to your server, you're all set, right? Well, that depends.

page you are viewing. This can be a great boon to beginners in how to achieve certain effects and designs with HTML. In fact, with many web-design products you can even choose to edit the pages you are viewing, which brings them straight into your editing software. This allows you to see them in both a design view and raw HTML code. Most people don't require a class to learn HTML, just time and a willingness to experiment.

From a design standpoint, making your own web site can often seem overwhelming. My advice is to create an outline of your web site, just as you were probably taught to do in school for an essay. Web sites are, after all, just structured bits of information. You will have a homepage, which is the first page that pops up whenever anyone hits your site, and linked pages below that. Each of those pages will have different elements, such as graphics and text. Outline what you think you'll want, and if you're adventuresome, maybe sketch out what you want the general look of your site to be. If you use the same graphics within your design for each page, the site will have a more unified feel to it, and this will also speed loading time for users of the site.

Once you have your page designed and posted to your server, you're all set, right? Well, that depends. If you've created the site simply for the enjoyment of you and your friends, then all you have to do is e-mail them the URL and sit back to accept their kudos. If you want more people to come to your site, however, you'll want to advertise.

With webcam-centric sites, there are two main ways to advertise. The first is to get your site listed on search engines. Most search engines have a link somewhere on their pages for those who wish to register their sites. They usually ask you to categorize your site so they know how to list it. There are even pay services that register your site with a great many search engines at once. If you're really looking to drive traffic to your site, this is an excellent way to start.

The second way to get your new on-line presence noticed is to register with an on-line webcam directory. An example of these is called www.webcamsearch.com, which lists webcams in categories of interest and region. Other similar sites are www.camcentral.com, www.earthcam.com, and www.cammunity.com. All of these sites have thousands of listings, and they're a great place to go to not only find webcam sites, but also to get ideas for your own.

Now that you've got your site listed, you can expect the visitors to start showing up. It's up to you whether or not they keep coming back. If you want people to make more than one quick trip to your site, you have to add content that gets updated often. The webcam is part of that, but other things include an on-line journal, lists of your favorite things, and links to other cool sites. We live in a world of channel-surfers, and it takes more than a simple text page saying, "Welcome to my homepage!" to keep most people's attention.

If you want people to make more than one quick trip to your site, you have to add content that gets updated often.

One sure way to get people's attention is to have streaming video, instead of merely a snapshot from your webcam that's updated every few minutes or so. And once again, with that magic word "video," the need for very high-speed bandwidth comes into play (we're talking multiple dedicated data lines). For this though, unless you have loads of spare cash just lying around, your web site will need to pull in money, which puts it into the class of business-related webcam sites discussed in the next chapter. (I just mentioned it so that, if someone mentions it to you, you don't feel bad you don't have it!)

Webcam sites can be a great way to keep in touch with friends and family, express yourself artistically, or just let the exhibitionist in you out for awhile. It's also a fun hobby and a great

Webcam sites may have a bad reputation with some, but like many of the applications for webcams that I've talked about, they can often help bridge the gap that communication through computers can cause.

excuse to learn HTML and web design. Whether you have one of those 3MB sites that your ISP provided when you signed up or you sprang for the whole domain name and hosted server, a web site can be like any other hobbyist obsession. There will always be something you want to try, something new you've seen that you'd like to try and replicate, and webcams can be an exciting part of that.

One way to get the benefits of having your own webcam site without the hassle is a service from www.securewebcam.com that I recently ran across. They provide you with a web page accessible only to you through a user ID and password. Software they provide takes images from your webcam and puts it on your web page. They have different levels of service in terms of how many old images are stored on-line, all with different costs. But recently they've offered free accounts to those who only wish to have 20 images on-line at any one time and who are willing to put up with a little on-screen advertising. This provides a neat solution to those who might just like to check in at home but don't want to deal with putting up their own webcam-enabled web site to do it.

Another alternative is called www.spotlife.com, which lets you post an updated cam image to the site, where cams are organized into categories. Unlike securewebcam.com, "Spotlife" is intended to allow you to share your images with others. This service is free, but only works with the Logitech QuickCam camera.

Webcam sites may have a bad reputation with some, but like many of the applications for webcams that I've talked about, they can often help bridge the gap that communication through computers can cause. After all, some studies say that the majority of communication between humans is of a visual nature. Humans say as much, if not more, with their eyes, hands, and gestures than they say to each other with words. All the applications I've talked about so far enable you to provide those visual cues. Video e-mail and videoconferencing allow you to provide those visual cues to a single person or sometimes as many as a dozen, but a webcam-based web site can give that information to the world.

By providing a window into their lives, the owners of many webcam sites—like many artists—provide a mirror in which to examine our own lives. Webcams are, for the most part (excluding staged adult entertainment), the truest reality-based programming available, and for whatever reason, people seem to need that these days.

It's definitely not for everybody, but whether you broadcast your life to five or fifty thousand, a web page-based webcam can be a tool for piece of mind to watch your children or your pets or an outlet for personal expression. However you use it, it certainly makes the Web a more interesting place to live.

Webcams are, for the most part, the truest reality-based programming available, and for whatever reason, people seem to need that these days.

PART 4:

Advanced Activities: In For a Penny, In For a Pound

Setting Up a Business Web Site

Chapter 12

Put your money where your homepage is

Businesses have distinctly different needs when it comes to designing and maintaining a professional web site.

Businesses have distinctly different needs when it comes to designing and maintaining a professional web site. One of the major hurdles might even be convincing the folks in upper management of the need for adding images from a webcam to your overall web design. Or you might need to convince yourself of the need for such an endeavor, and decide exactly how much of your resources in time, capital, and personnel should be devoted to the project.

Assuming you decide to go ahead, the next major obstacle is security. Many companies are under an information-technology mandate not to deploy any new systems unless they are secure. And even companies operating without such a mandate probably aren't going to be happy if the addition of a webcam application means hackers can steal confidential data stored on the same server.

Beyond the security problems, there are many legal issues surrounding web content that have been worked out in recent years and some that are ongoing. The on-line world is less like the Wild West of yesteryear and more like the frontier settlement towns that followed, but on-line law is still somewhat sketchy. There are established rules that businesses need to follow concerning content though, and the addition of the webcam makes this area much more complicated, especially if you end up filming people on private property with or without their knowledge.

Once these major hurdles are overcome, there are two more minor issues that a business will have do deal with. These last two are the concept of how the site will be designed and how the content on the site will be maintained. These are not show-stopping issues like legal and security concerns. In fact, maintaining content and designing the web site are issues faced by consumers and businesses alike, differing only in scale for the most part. If you fail in these two areas, you won't get sued and you won't have your company data stolen, but you won't accomplish your stated goals either.

People need to reason out exactly what they're trying to accomplish with their webcam.

Establishing a business case

There are generally two reasons people in business have to establish a business case before they make the jump to implementing a web camera as part of their web site. The first reason is that people need to reason out exactly what they're trying to accomplish with their webcam, and then figure out if their ideal situation is both technically and economically feasible.

A lot of times, the strongest supporters of a web camera addition to their web pages will discover that there are better ways to achieve their goals without the painful experience of learning this after the fact. Or you might discover that there are not enough people in the office to maintain the camera. If you really feel that the camera can bring in additional revenue greater than the associated cost, you can always throw money at the problem, but you'd better plan this out before you invest in a money pit.

Even if the rank-and-file employees feel that a webcam application is the best way to go, upper management may not think the investment is important.

The second reason that a business plan might be necessary is that, even if the rank-and-file employees feel that a webcam application is the best way to go, upper management may not think the investment is important. But presented with a detailed plan that takes cost into consideration, few managers will abandon a plan simply out of their own prejudices or technological fears.

A very affluent chief information officer for the federal government once told me, "The easiest way to get something done is to know what needs to be done." He advocated business cases for most every technical improvement or deployment in the entire agency, and his agency was consistently ranked higher than other large government bodies each time Congress decided to conduct a technological spot check.

And the business plan does not need to be complicated. One of the simplest components to the plan, which often is the entire plan itself, is the "Wants/Advantages/Disadvantages chart" (see Figure 12-1). This is simply a three-column chart where the people making the plan put their desires for the webcam in the first column and the advantages and disadvantages that will occur if this goal is achieved,

Wants	Advantages	Disadvantages
1. More website traffic	1. Allows us to charge more to advertisers	1. Costs more to maintain site - Cost -Need to update content more often
2. Pull traffic from competitor's web site.	2. Our website is a marketing tool we control	2. Difficult to attract visitors in crowded online marketplace
3. Give website unique "personality"	3. Can lead to more online sales	3. Additional hardware and software expense, training of personnel
	4. May add new demographic of buyers to consumer base, increasing marketshare	
	5. Buyers who visit our site are more likely to buy from us again	
	6. Causes greater brand awareness and web presence	

Figure 12-1. Sample "Wants/ Advantages/ Disadvantages chart" for setting up a webcam at the office.

making sure to consider resource cost in each line (normally in the disadvantages column).

If you do this and can't fill in the "wants" column, then you might be barking up the wrong tree, so decide up-front what your goals are for the camera. Do you want to show a beautiful location to increase tourism in your area? Do you want to prove that handmade products on your web site are legitimate by displaying the artists at work? Or does your company have something so interesting going on that you feel showing this event occurring live would drive a lot of traffic to your site and increase sales? Whatever your reasons, just be sure you know what you want, or else you won't ever get where you need to go.

Another program-ending event is if your disadvantages far outweigh your advantages, meaning you might need to consider a better solution. But if you discover that your webcam idea accomplishes all your goals with acceptable trade-offs, then the simple chart can be spruced up and delivered to the boss as a graphical argument for your idea.

Security

There are various methods to make sure that people stay where they are supposed to on your web site.

Security is the single greatest concern when using the Web for most companies, whether as an information-delivery system or a full-scale e-commerce application.

In its simplest form, a web page is a window shining into your company's network that can be accessed by anyone anywhere in the world. In a way it's like having a picture window facing a major highway. It basically invites people to gaze into a small part of your home, which in this case is your company LAN (Local Area Network). Most companies make this area as pretty as possible in an attempt to show their organization in a good light. Nobody wants a messy front room if it's known that lots of people are going to be looking inside.

A webcam by its very nature has to be open to the public, much like a traditional web site. But a webcam offers some security problems, since it shows physical details of your office—live in some cases—and because the increased amount of traffic makes it easier for hackers to disguise their actions. After all, many users will hang out watching your camera for a long time, so determining which user took the time to wreak havoc on your network is that much more difficult.

There are various methods to make sure that people stay where they are supposed to on your web site. The most common is the firewall. A firewall is basically a dedicated device or program that is configured to know which activities or actions are allowed on your network and which are banned.

A well-programmed firewall can make your system very secure, keeping everyone (who comes in via the front door to look at your webcam) from accessing the rest of the house.

Perhaps certain users are allowed to come in from the front room and wander through the rest of the house, or perhaps all authorized users access the LAN from a different portal. Some types of activity might be banned all together, such as uploading data or overwriting files on the server. A well-programmed firewall can make your system very secure, keeping everyone (who comes in via the front door to look at your webcam) from accessing the rest of the house.

The problem is that programming a firewall is a constantly changing and difficult task to accomplish. For example, say an administrator uses the firewall to block all the active paths from the webcam area of the LAN to the rest of the network. That may work one day, but as soon as new users get added to the network, or the network topology changes even slightly, new holes can occur.

And a poorly configured firewall can do more harm than good. This was readily apparent at an office I worked at where the director of information technology decided to build a firewall on a network that previously had no security whatsoever. This was a noble goal, but poorly implemented.

What this person did was to block all incoming TCP/IP packages sent into the network altogether. This made the network secure by blocking people from entering it, but also made the entire LAN unusable, since connected users could no longer access the Web, which was a major part of my job at the time.

What happens when a person types in a web address is that the www address or URL (Uniform Resource Locator) gets sent to the DNS servers, which is basically a huge database of what the letters in a web address actually mean. At the DNS server, the URL gets translated into a series of numbers that tell your computer where to go to find the site. So you might type www.gameindustry.com, but following the bounce through DNS your computer sees 199.240.78.84 or some other addressing number. Try typing in the number and see what happens. But unless you know the numbers for each web site, you need DNS servers.

Since that translation needs to be sent back to your computer once the DNS looks it up, that data was hitting the newly installed firewall. The end result is that all users were being told that CNN and *The Washington Post* on-line sites, or anything else you could type in, could no longer be found. We sure had a secure network then. Even authorized users were turned away.

With higher-end operating systems especially, the sheer number of services running on most systems offers loopholes to the experienced hacker.

But as long as IT directors know what they're doing, most firewalls offer good protection for networks and should be required for any business web server. This is not to say that it's a 100 percent safe solution. As with any type of security, there's always the possibility it will be breached. Especially with higher-end operating systems such as Microsoft Windows NT, the sheer number of services running on most systems offers loopholes to the experienced hacker. I'm no hacker, but I can think of at least four exploitable services that can be used to get around a firewall. If your IT administrators know these, they can manually plug the holes on the firewall, but the point is that security is a constant battle.

If even the best firewall, configured perfectly, is not a 100 percent guarantee of safety, what can you do and what should you watch for?

Hackers are always trying new things, and once in a while these new methods work. Normally, this means a hacker will be able to break into a site and change the content. This is embarrassing but not really dangerous, unless the company doesn't keep backups of its web pages close at hand—and out of reach of the on-line community. The best defense besides a firewall is to make sure you have the latest updates and patches installed for your operating system and applications. These are almost always available from the software company's web site and usually include security fixes, both major and minor. They also often include explanations on what security holes they are designed to close. For more information, study Chapter 15, which deals with security issues in more depth.

There are several programs out there that spoof networks, making hackers think they've broken in when in fact they're wandering around a ghost network.

The government is often a target of attacks from hackers. While most are generally amateurish in nature, government agencies such as the Air Force, NASA, and even the Department of Justice have fallen victim to hackers (see Figure 12-2).

Since they are such inviting targets, some government agencies have come up with a foolproof method of keeping their LANs safe from intruders: they've simply disconnected their web servers from the rest of their networks (see Figure 12-3). That way, if hackers manage to break through a firewall, they won't be able to do anything other than post their own content or format the drives in the machine. Since backups are created each day—hourly, in some cases—the original system and its content can be up and running again minutes after the hack is detected. This is a relatively simple solution that most large corporations also have implemented, and midsize and smaller companies could easily implement it as well.

Cutting-edge agencies have even made it easier for hackers to break through firewalls into computers that are not connected to the LAN. There are several programs out there that spoof networks, making hackers think they've broken in when in fact they're wandering around a ghost network. And guess what? They're going to be in for a surprise, because in most cases any intrusion into a ghost network is

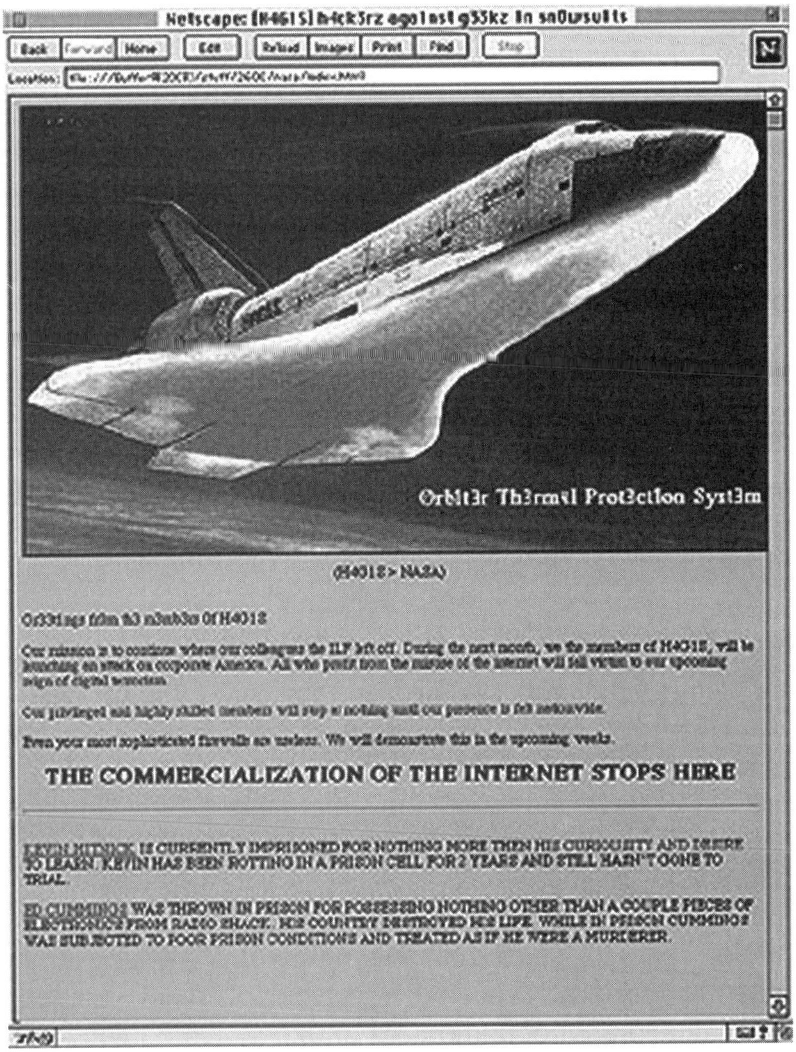

Figure 12-2. Screen shot of a hacked NASA web site. The government is often a target of attacks from hackers.

immediately logged. Sometimes a rapid computer-response team is called that watches hackers and traces their origin. By the time hackers decide to leave or learn that they're actually sitting in a clever trap, law enforcement is either en route to their location or their profile has been sent to the appropriate agency so a case can be built.

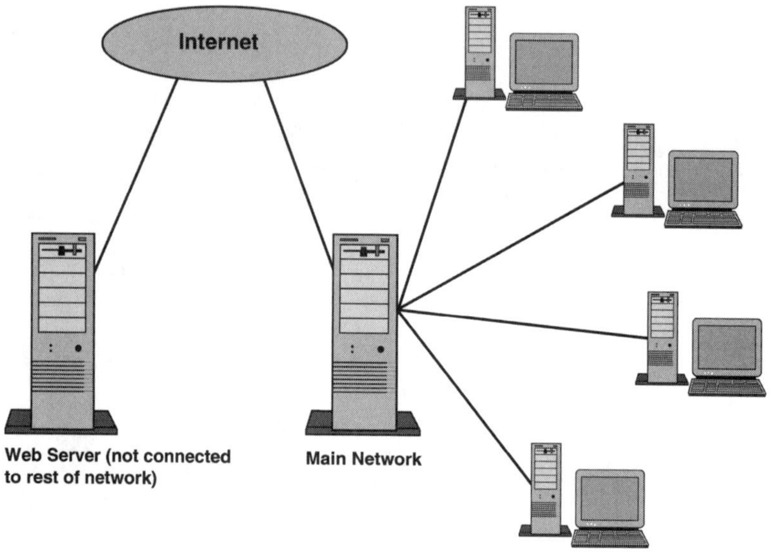

Figure 12-3. Diagram of a web server disconnected from LAN. This is a foolproof way for companies to keep their networks safe.

But unless your company has a lot of enemies, simply disconnecting your web server from the rest of your LAN is probably all you need to do to protect confidential data. You can also install a firewall on the orphaned server, but again, for most companies even this might be a bit of overkill (though some software-based filtering of TCP/IP traffic is always a good idea).

Another security issue that doesn't involve hacking is the concept of sending live data from your place of business. Say you have a lot of computers or expensive equipment in the room where you want to put your webcam. And say a thief logs on to look at your site. Most likely the thief can find your company address somewhere on your site. Then he or she can simply watch the camera to see when people are in the office and when the office is empty.

It doesn't take long for someone to discover that nobody is in the office after hours, or that the place is deserted over the weekend. This sets up a nice "smash-and-grab" opportunity for thieves who want to break in and get whatever they can. Ironically, many businesses spend so much time with Internet security issues surrounding their new webcam that they forget it also punches a hole to some degree in their physical security. Of course, if you're shining the webcam outside (as in the case of a tourism bureau), or into a fish tank or something, this is less of an issue or not an issue at all.

If you have a camera inside a typical office, however, most companies that have successful webcams either use nondescript internal rooms where an outsider would not know how to get to the area on camera, or they simply limit the hours that a camera is in operation. The camera's image can simply be programmed to freeze a frame around 5 P.M.. and not resume live filming until the next working day. Or the system can default to the emergency image used during crashes, such as a company logo or a still image of a particularly interesting event that occurred previously. After all, showing a darkened office filled with high-tech gadgets is not only an invitation to theft, it's also very boring.

Just be sure to make it clear to your users what hours the camera is in operation, so people don't think the camera is never actually activated or that you're trying to fake live footage with preshot images.

Legal issues

There are very few laws that specifically pertain to webcam use.

There are very few laws that specifically pertain to webcam use, which is not to say that you can use the device with impunity. It's just that when a company gets in trouble legally for its web site content, it's often not the camera's fault.

The main thing you need to worry about is capturing people's images for commercial use without their permission. You're covered if your webcam is sitting in a public place and taking pictures of another person who is in a pubic place, though this is probably a rarity when dealing with webcams. Still, know that if you are in this situation, that you can basically take images of anything you want. If a naked hostage comes running out of a building onto the sidewalk, and you catch that person on your webcam, you are well within your legal rights to do so.

Although it hardly seems fair, the other person gives up his or her legal rights to privacy the second that person steps out into a public place. This is the rule taught to newspaper photographers on their first day on the job.

But more likely you're going to have your webcam in a hallway or specialized room in your office, perhaps the conference room. This presents a bit of a dilemma, since both the camera and the subject are in a private place. Most likely, if you have the camera in a hallway, then you will be fine in court (see Figure 12-4). If you have it hidden in the bathroom, then you could be in a lot of trouble.

One way to make yourself near bulletproof legally is to have all your employees sign a statement saying that they understand they could be occasionally filmed by the webcam.

One way to make yourself near bulletproof is to have all your employees sign a statement saying that they understand they could be occasionally filmed by the webcam or cameras at location X, Y, and Z in the building. When visitors come, you can advise them of the same thing, though it might be a bit much to make them sign a similar form. Basically, make everyone aware that the cameras are there and that images are being uploaded to the Web live, or nearly live.

Where you will more likely get in trouble is the general content of the web page around the camera. Be advised that when you publish to the Web, you are indeed publishing data

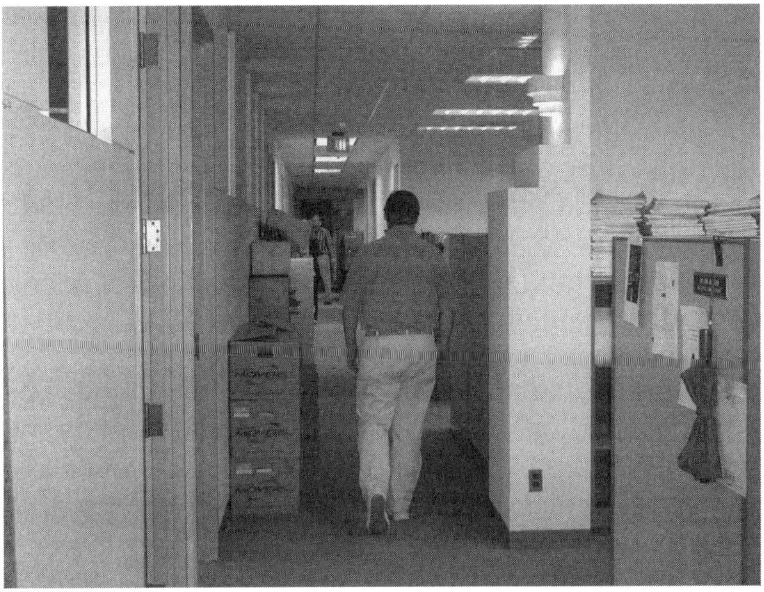

Figure 12-4. Placing a webcam in an office hallway probably won't get you into any legal trouble.

and can be held responsible if you slander or commit libel against someone. That means if you print something that is not true about a person, or even a group, then you could end up paying for the mistake if someone sues.

Designing the site

When setting up a web site, there are many considerations that go into the puzzle. But when you add a bandwidth-intensive webcam, many design techniques are elevated from mere tips to critical needs.

A prime example of this is the use of graphics on a web page. Of course, in the graphically intensive World Wide Web, a site without graphics is not going to be very popular. If users go to a site and see text and nothing else, they probably aren't going to stay very long and most likely won't be coming back. But a lot of web pages go to the opposite extreme. They design their pages with hundreds of images, or a few very large, high-resolution images, which can be just as bad (see Figure 12-5). Sure the page will look good, or at least could look good assuming that a competent artist did the work, but if the page takes too long to load, few people are going to hang around long enough to see it in its full glory. This is a common design rule on the Web.

Most design programs will even estimate how long the page will take to load on a variety of connections, such as a 56K modem. But as soon as you add a webcam to your site, the entire demographics will change. A webcam is the equivalent of a large graphic, so other graphics will have to be eliminated or reduced if you have any hope that the average person will see the entire page quickly.

And that's just if your camera is broadcasting an updated image every so often. If you have a live camera delivering

Figure 12-5. Web site with a lot of graphics, meaning that the page will take a while to load. Adding a streaming webcam would make it too slow for normal use.

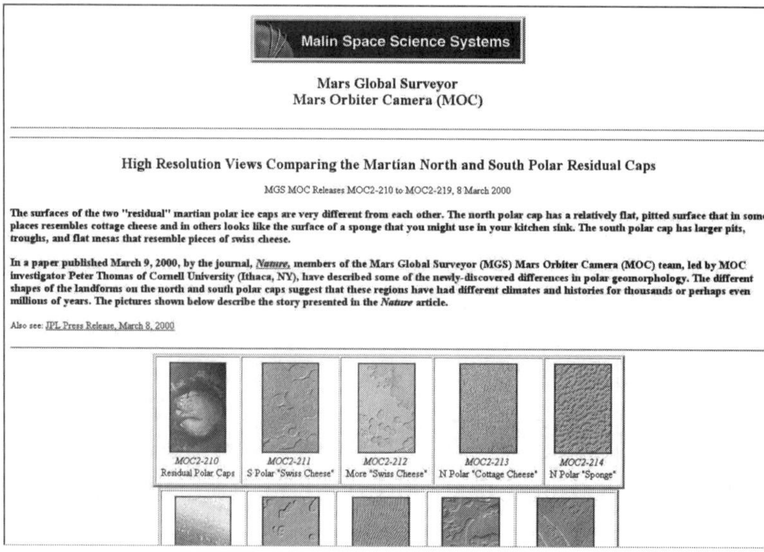

streaming content, then the load time will increase dramatically. A streaming image can easily overload most standard modems to the point where the user has to wait a long time to get images, and adding an excessive amount of graphics to the page in addition to this will be harmful.

First off, the extra graphics are going to be competing with the camera for bandwidth to the user, meaning it may take a long time for the user to begin getting the video feed. And secondly, the process of downloading the images could possibly interfere with the transmission itself, which will lead to lost packets of data and possibly stop portions of the page from appearing.

Running a webcam over the Internet is difficult enough in terms of bandwidth, so make sure you aren't competing against yourself with large graphics on your own page. A live image will hopefully be interesting enough to almost stand on its own, and thus doesn't need a lot of complex graphics backing it up anyway.

A second mistake companies with webcams make is the use of sound on the same page—or even the same server—as the webcam.

A second mistake companies with webcams make is the use of sound on the same page—or even the same server—as the webcam. You might think that it would be cool to show your company off to visitors while some patriotic music plays in the background. And in an ideal setting, you might be right. But in terms of bandwidth, just because you can't see it doesn't mean that it is not there. Each bit of sound you put on the page either has to be transmitted to the user, much like streaming video, or downloaded and then played on the user's computer. If you choose to push the music to the user's computer, most browsers will warn them that a file is being downloaded to the system and ask them if they want to allow this to happen. This prevents people from sending a malicious program to people who visit a web site, or at least it makes it harder. This extra step

Remember the problems of broadcasting streaming media to one person? Well, those same problems can be ten times worse when you have hundreds or thousands of simultaneous users.

is more than enough to annoy a user, and suspicious people will wonder why your company is trying to implant a program on their computers.

But even if you are transmitting the sound, you are using bandwidth. This is the main reason most webcam sites don't have sound. Since you're already pushing a heavy stream of data to the user, the addition of sound can really hurt, whether it's sound from the video feed itself or background music. And remember that you are potentially broadcasting to a lot of people, which is different from doing a videoconference.

Remember the problems of broadcasting streaming media to one person? Well, those same problems can be ten times worse when you have hundreds or thousands of simultaneous users. And while poor quality might be acceptable for a personal web page, you don't want to base your business on it.

The last potential pitfall is the use of frames on your page (see Figure 12-6). Frames is a technique used to put multiple scroll bars and multiple windows on a web page for the purpose of better organization—and easier programming, from a designer's standpoint. While most every browser in use today can accept frames, the problem is that they take longer to load, and each frame potentially has its own graphics that will again compete with the webcam.

You can still use frames, if you are very careful. But test your page on an end-user's computer to see how long it takes to load and to get webcam images. If it's more than a few seconds, kill the frames. Also, be aware that some browsers display frames differently, so test your page with all the major browsers to ensure that the images and frames are displayed correctly.

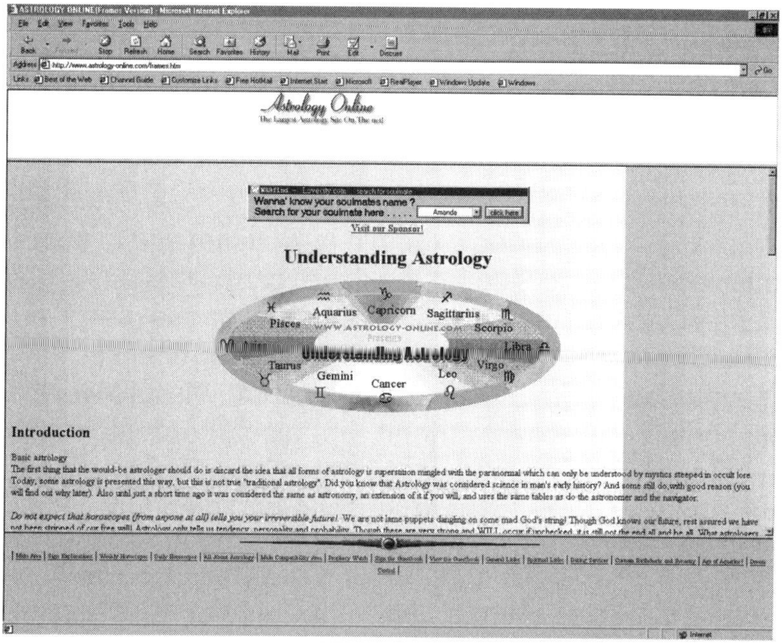

Figure 12-6. Web page with a lot of complex frames, which also take a long time to load.

I'll make you a star, baby

Finally, it all comes down to maintaining content. A web page that has a lot of good content—be it articles, or information, or just interesting things to do—is going to succeed, whereas a boring page won't.

You are ahead of the game on this one with the addition of the webcam to your site, since the image is always fresh, or at least near fresh. Since the camera is always taking images, either in real-time or every few minutes or hours, you won't have to worry about updating content. It will update itself.

But don't ignore the rest of your page either. Used most effectively, the webcam site is a perfect blend of interest—

generated by the camera—and information, which is generally used in a support role. The best company webcam site will bring users in with the webcam, then take the opportunity to let visitors know about products and services that are sold by the host.

In the early days, a good example of webcam use to support a business was the Dupree the lizard cam, who was a giant iguana and company mascot for Iguana Images (see Figure 12-7). Dupree had thousands of fans that checked in on her everyday at http://iguana.images.com/. Visitors were not only shown their scaled friend, but also information about the company that "employed" her. I visited Dupree all the time, and was sorry to learn that she had died following an eight-year run as a true Internet star.

On the Dupree memorial site, the company has statistics posted about usage when Dupree was alive. Links to the site came in from around the world. And this is just one example of how a webcam can boost traffic to your pages.

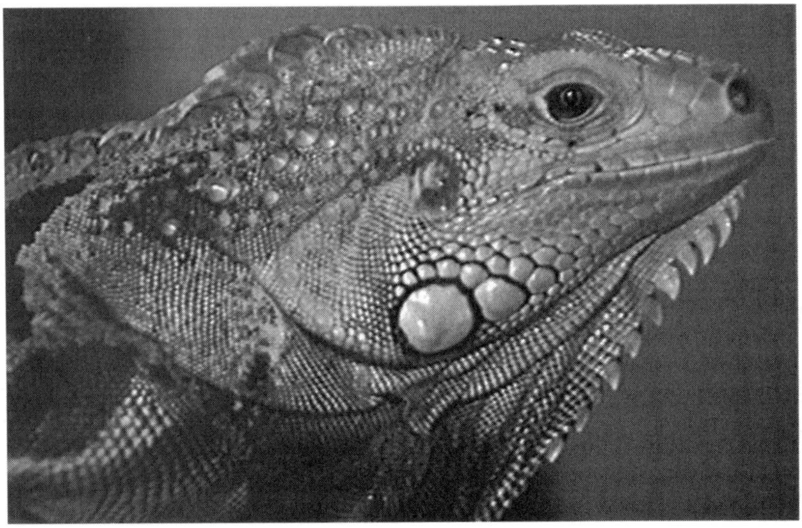

Figure 12-7. A good example of webcam use to support a business was the Dupree the lizard cam.

Last, but not least, be ready to grow. If you do a good job of managing content and bringing people in, you may quickly find your company web site (and hopefully your company!) more popular than you ever imagined. While this is a good position to be in, all that sudden popularity can be wasted if you're turning people away due to a slow connection.

Just because your current web site has enough hardware and bandwidth to handle your users does not mean that a clever webcam won't double or even triple the usage. And when the world beats a path to your door, you'd better make sure you have a large welcome mat. Web users are fickle by nature, and you rarely get a second chance to impress someone turned away by poor performance.

As part of your overall plan, make sure you factor in the costs of upgrading both your hardware and your connection to the world. Contact your ISP and see how quickly it could upgrade you in an emergency, tell your IT staff to monitor web traffic, and be ready to install and purchase new servers if needed.

The Internet is kind of like a small town in that news travels fast by word of mouth—or in this case, e-mail. Couple that with a story about your new site in a few e-zines or even traditional news outlets, and you could find your company's popularity speeding ahead like a runaway train. Just be sure you are ready to get on board, or success can crash you just as quickly as failure.

If you do a good job of managing content and bringing people in, you may quickly find your company web site (and hopefully your company!) more popular than you ever imagined.

Special Webcams: Thinking Outside the Box

Chapter 13

There's a whole range of devices out there that can be used as webcams.

I've mentioned them in passing throughout this book, but I thought I'd take a moment to talk about some of the nontraditional webcams out there. By nontraditional I mean those that aren't simply a small digital video camera connected directly to your computer via a USB connection or other means.

There's a whole range of devices out there that can be used as webcams. Basically, almost anything that can provide an image under Windows for example can be used as a webcam. This includes video cameras with video-capture cards, digital video recorders, and digital still cameras. It also includes specialized digital video cameras that are stand-alone units, using either a wired network connection or a wireless connection to send images.

If you plan on having more than one webcam in your house, often the easiest way to do it is to buy these stand-alone cameras and a network hub. It may sound expensive, but it's usually quite a bit cheaper than having both computers and webcams throughout the house, and you'd have to network the computers together with a hub anyway.

There are even remote cams that sit on a movable housing, allowing you to rotate and tilt the camera for the best angle through a network interface. For school classrooms and businesses, these specialized cameras can allow people to do things they simply couldn't do otherwise. A wireless cam shielded in a weatherproof casing can broadcast images of wildlife to children involved in a class project, or a rooftop cam could help a delivery company monitor the flow of traffic around its business.

Using traditional video cameras also offers benefits such as better image quality, infrared video, and, of course, an analog backup in the form of a video cassette if you choose.

Using traditional video cameras also offers benefits such as better image quality, infrared video, and, of course, an analog backup in the form of a video cassette if you choose. I've even heard of some companies offering both a videotape and a live Internet broadcast of events such as weddings and meetings. They can do this simply by using a video camera hooked up to a video-capture card and a computer connected to the Internet.

I also know of many researchers using webcams to cheaply and easily study animals, weather, and other natural events.

Mobile webcams have also become more popular lately. Some notebooks now come with a camera embedded just above the LCD screen, which is great for presentations and videoconferences. You can even get cams for handheld devices such as the PalmPilot.

While I can't detail every type of webcam and its use, I did want to mention some of the possibilities for them. Oftentimes, it's simply a matter of going on-line and finding the manufacturers that make the kind of webcam you're looking for. In addition, many on-line newsgroups and web sites have users with years of experience who can help you find the answer to a unique situation or problem.

Due to the basic simplicity of the technology, as well as to advancements within it, webcams are becoming increasingly common and are providing answers to uncommon problems.

Webcams are becoming increasingly common and are providing answers to uncommon problems.

PART 5:

Things to Consider and Food For Thought

Tool or Toy? The Corporate Question

Chapter 14

People often wonder whether the technology can really be effective at work.

I decided to include a brief chapter on webcam use and its overall place in a business environment, since people often wonder whether the technology can really be effective at work. Webcams have grown in popularity over the years, but more so from the new brand of Internet stars than from business success stories.

These pioneers that have advanced the field have both helped and hurt the market to some extent. While they have helped to make webcams standard on some desktop computers and even notebook computers, mostly they are still aimed at the home user. Most of these Internet stars, many of whom have cameras throughout their homes that are on all day, are flamboyant to some extent. Some even have cameras in their bedrooms. And this heightens awareness on the consumer side, while turning off to some extent businesses that would prefer not to be associated with their antics.

But make no mistake: some businesses are using webcams effectively. It all depends on how you make use of the technology. If you think about it, almost anything can be considered both a toy and a tool depending on how it is used. Take a car, for example. The car can be used to take you to your work, to the store, or off on long sales calls. Used this way, the car becomes an invaluable tool and a near necessity of modern life.

But a car can also be used on leisurely Sunday drives. It can be used to race other people with cars. And the backseat of some cars has long been used as a secluded hideaway, depending on where you park, for young lovers. Used in this way, the car falls squarely on the toy side of the question. It's fun to have and brings pleasure to life, but you could live without it if you had to.

It's a lot harder to give up tools. As time passes and bandwidth increases, I think you will see webcams become more of a tool for businesses. Look at technology in general. In the 1980s, technology helped businesses compete, but for the most part in a secondary role. Then in the 1990s, technology was known as an enabler for businesses, and without it they could hardly keep up. Today, technology practically is the business, whatever that business may be.

Make no mistake: some businesses are using webcams effectively. It all depends on how you make use of the technology.

Personally, I got into covering webcams in the early days, when few people even knew what one looked like. And truly, there have been more fun consumer successes than there have been Harvard business case-study material. But I know of several innovative business uses of webcams, some of which have been quite profitable. I want to describe a few to get your mind thinking about your own business and where you might be able to fit one into your marketing strategy.

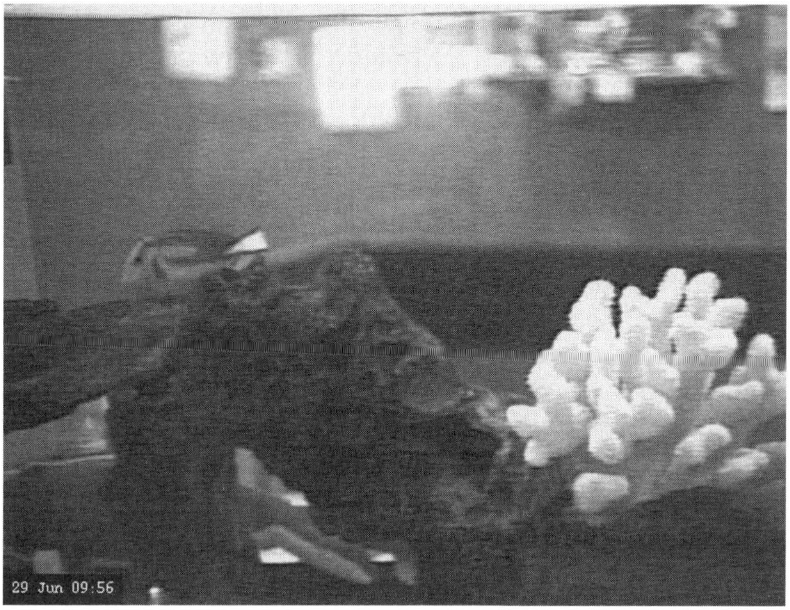

Figure 14-1. A dentist put a webcam facing the fish tank in his office lobby, and soon people were visiting his homepage. His Internet goldfish net him new customers.

The first thing to consider is that many of these webcam-consumer sites are, in fact, businesses. Many charge people to see live content or show updated pictures slowly to nonpaying guests and very quickly to those with premium access. Some also sell ads off their site, on the theory that with so many people visiting, there's a large potential market for advertisers.

Needless to say, the porn industry has capitalized on this. Ever intelligent about how to make a quick buck, the industry has taken the idea of voyeurism to new levels, but I don't think you should try this as a way to increase your business.

That brings up the first potential use of a webcam for a business. If your business is focused on bringing people into your pages, there are few tools as good as a webcam. A dentist in a busy city put a webcam facing the fish tank in his office lobby, mostly as a novelty (see Figure 14-1). But soon people were visiting his homepage, and those that lived in the same area wanted to visit the "fish dentist." His collection of guppies and goldfish ended up netting him new customers.

If your web page is driven by advertising, then so much the better. This way, you don't have to wait for secondary benefits, such as new customers visiting your office. More traffic on your site translates to more hits—and more advertising revenue, if your sales staff stays on the ball. Just be sure that your webcam subject is interesting, or at least unusual.

More traffic on your site translates to more hits—and more advertising revenue, if your sales staff stays on the ball.

Another use for webcams in business is what I call the business-enhancement model, where the camera becomes a true tool, often for your customers. Day-care centers are the primary users of this model, but I have also seen this done at a few kennels and pet-care places, which, I suppose, are a form of day-care as well.

What happens in this model is that the camera is set up so that parents or pet owners can look in from their home or office and keep tabs on how well their children or pets are doing. Businesses that boast the ability to let customers see their children around the clock—assuming the children are in the room with the camera—can use this as a way to get a leg up over their competitors who don't have this ability. If you decide to go this route, you should consider making the site password protected, so that only users with children at your school can view them. Otherwise, you could be doing more harm than good.

I have also seen this model used in restaurants in certain high-tech cities such as San Francisco. The camera is set up so that people can see how crowded the place is at any one time. This works great, because the few people who might decide that the place is too packed are made up for with the people who see that the place is empty on off-hours and decide to come in for a bite. And the time when you want people to come in is when your place is emptier anyway.

Another role for webcams might be what I call the secondary role, where the public or the customers are not generally the ones that make use of the camera. This includes the more common videoconferencing cameras as well as the more obscure uses such as monitoring a construction site from a remote location.

In this case, the camera can serve to make you more productive by either saving time, saving money, or supervising quality. It doesn't make any money for your business by itself, but it can put some more income into your bottom line.

Businesses in general are somewhat slow to deploy webcams, because the bottom line is not easily definable. You can compute easily the fact that the camera saves money by letting you chat with other executives for free over the Internet, but most people don't realize what the savings are over a year. And a penny saved in business really is a penny earned.

There are probably 10 ways any business in the world could deploy a webcam in some role and use it to make money, or at least save some. My bet is that, if you sit down and ponder every aspect of your current business, you could come up with something.

There are probably 10 ways any business in the world could deploy a webcam in some role and use it to make money, or at least save some.

Almost anything can be classified as a tool or a toy based on how it is used. Even computers, the mainstays of business today, have a tremendous amount of potential either way. Just be sure everyone at your office or business understands that you want to use the business webcam in a tool mode.

Security: Protecting Yourself and Your Computer

Chapter 15

Webcams offer special challenges when it comes to security.

Security should be a concern of anyone who sends data over the Internet, as well as anyone who uses a computer to store personal information. In short, security really is everyone's concern. You can often feel a little overwhelmed when thinking about all the things that can go wrong where security and your computer are concerned. The secret is to develop a plan and take it step-by-step. There really isn't anything that tricky to keeping your data safe, just common sense and taking advantage of the tools available.

Webcams offer special challenges when it comes to security. A webcam isn't very much fun if you can't use it to share pictures and videos with people, but doing that can increase your risk. The following checklist outlines what I believe to

Table 15-1.
A checklist of common sense security issues.

The Common Sense Security Checklist

- ***Raise your drawbridge:*** If possible, make sure your network connections or computer are shut down when you are not using them. This will greatly reduce the chances of anyone hacking your computer.
- ***Guard your borders:*** Invest in an anti-virus program and a software-based firewall. (Business users might want to invest in a hardware firewall, if they don't already have one.)
- ***Loose lips sink ships:*** Don't give your password or user-IDs out to anyone.
- ***Friend or foe?:*** Don't download or run applications from strangers or unknown sources.
- ***The bigger you are, the harder you fall:*** The higher your profile on the Internet, the more likely it becomes that you will be targeted by a hacker.
- ***Look every gift horse in the mouth:*** If something is too good to be true, it probably is. Scams that promise money or other gains often can camouflage attempts to gain access to your computer or personal information.
- ***Know thyself:*** Know beforehand what you're going to use your computer for and what possible security holes these activities might open.
- ***Never assume:*** Not to sound paranoid, but most security problems come from someone thinking, "Well, that will never happen." Complex computer applications and operating systems leave a lot of leeway for mistakes that hackers can take advantage of.
- ***Looking out for No. 1:*** ISPs and other on-line companies try to protect their customers the best they can. Yet protecting your personal data and computer is something that can't be left to others. Take an active role in discovering and understanding how and where you are covered.
- ***Finding the path through the wilderness:*** The Internet is still in many ways frontier country. There are paths through it, and there are lots of things to explore, but remember that the farther you stray from the path, the more likely you are to run into trouble.

be the essentials of security for webcam users, or anyone who has reason to share lots of data over the Web (see Table 15-1). Now that I've taken all the fun out of the Internet for you, let's take a more detailed look at the subjects outlined in the checklist.

So what about these Internet connections that advertise that they're "on" all the time? Does this put you and your computer in greater danger? You bet!

Raise your drawbridge

With a 56K modem, your computer is completely secure from security hacks when you're not dialed in. That unconnected modem is an impenetrable wall to any hacker. So what about these connections (such as cable modems and DSL) that advertise that they're "on" or connected all the time? Does this put you and your computer in greater danger?

The answer, in case you haven't already guessed, is: "You bet!" Think of that always-connected line as a window into your computer. It may take a bit of tinkering to get it open, or a big enough rock, but sooner or later someone will figure out a way in. To extend the window metaphor just a little more, hackers may not even need to "break in," as it were; instead, they might be able to look through the window and see what you have inside.

So how do you replace that connectivity window with something like a door, or even better, a wall? Well, as I said, the best advice is to shut off your computer when not using it. Also, depending on how your connection is set up, you may be able to disconnect your computer from the outside line with a simple click of the mouse. If you need to leave your computer on and connected all the time, or still just prefer to, then you'll want to consider the next section about how to . . .

Think of a firewall as your computer's very own border guard. In a sense, it asks the data where it came from, where it's going, and what it's planning on doing.

Guard your borders

You probably have heard of anti-virus software. This software looks for actual virus code or virus-like behavior and helps you guard against it. This is an absolute must for anyone who plans to do anything on the Internet. Viruses are the most common type of computer security problem and can wreak havoc on your computer, as well as pass along sensitive personal information to hackers.

Most anti-virus software today contains some kind of agent that runs all the time and monitors all the files and programs you open, run, or view to make sure they are clean. This is the best type of protection, as it can actually prevent a virus from becoming active on your computer.

Now, the next term I'm going to throw at you is a little less talked about, especially for home users. The term is "firewall," and it's used to describe either software or hardware that monitors the data going across your connection. It doesn't really look for viruses; it looks for attempts to hack your computer. It does this by looking at the type of data coming across your connection and seeing if it's the right kind of data and if it's going to an allowed area.

Think of a firewall as your computer's very own border guard. In a sense, it asks the data where it came from, where it's going, and what it's planning on doing. If the data's "passport" is accepted, then it is allowed on its way. If the firewall detects something is not right, however, then it simply will not let the data through. This allows the flow of normal traffic to continue unimpeded, but stops hackers from digging around to find a way in.

Firewalls can be configured to fit your needs, and so you can shut off some entry points completely, have the guard check data at others, and totally open other entry points. Most home users who are interested can probably pick up one of the many new personal firewall products on the market. These are software-based and provide the rudimentary versions of these functions for your personal computer. Most of them come preconfigured to handle the most common setups and can be modified to some degree.

Both the firewalls and the anti-virus software help make your connection more of a door than a window. Good data traffic is let through, harmful data is kept out, and prying eyes are forced to look elsewhere. But don't think this makes you or your computer invulnerable, because as everyone knows . . .

Loose lips sink ships

All sorts of security problems can arise from the simplest of mistakes.

All sorts of security problems can arise from the simplest of mistakes. For example, you might receive an e-mail from your ISP saying that it needs to have your user ID, password, or credit card number to verify something, or to give you an upgrade, or any of a thousand other excuses. Your ISP asks if you would be so kind as to send that information in an e-mail, which, for arguments sake, let's say you do. The next thing you know, your e-mail has been hacked, you can't log on, and your credit-card bill is a few thousand bucks higher.

The facet of this hypothetical situation you did not consider was that your ISP has your phone number, and if there were a problem with your account, it most likely would send you an e-mail asking you to call the customer service number. While some of you may scoff at the thought of anyone falling for this scam, it has happened thousands of times over the years at America Online alone.

Friend or foe?

One of the biggest problems with the Internet is the fact that so often it can be difficult to tell friend from foe. There are viruses out there (some of them quite famous, such as "Melissa" and "LoveBug") that can replicate themselves by going through your on-line address book and mailing copies of themselves to people you know. So even an e-mail from someone you know can cause all sorts of problems. The best advice is to not run or open any file attachments unless you know for sure what it is, and even then be cautious. The best protection is a recently updated anti-virus program and a cautious attitude.

If you're using a webcam to broadcast an image to a web site or using it to chat with people, then be aware that a cam can give away information simply by being on. It can tell people when you're alone or not at home; it can show what kinds of things you own and other information that could be used to harm you or your possessions.

The best advice is to realize that the Internet is just like any other public place, only with the added danger of you not being able to "see" everyone who can see you.

If you have a web site or video chat with strangers, your profile is certainly going to increase, which means your risk will as well.

The bigger you are, the harder you fall

Speaking of being seen, that leads us to our next topic, which is profile. If you only use your webcam to talk with relatives and friends, you aren't likely to run across any problems. If you have a web site or video chat with strangers, however, your profile is certainly going to increase, which means your risk will as well. As mentioned before, the progenitor of the personal webcam, Jennifer Ringley (a.k.a Jenni from Jennicam), has to deal with thousands of attempted hacks each week. Of course, her situation is at one end of the spectrum, but hers is also one of the highest profile sites on the Internet.

I'm not going to tell you not to have a web site or not to take advantage of your webcam and talk with people from around the world. My point is simply that there is a higher risk involved, and realizing that is a big step toward stopping any problems from developing.

The old adages are the best security advice you can have in the modern Internet-connected world.

Look every gift horse in the mouth

Humans are a gullible lot. We always seem to believe that we as individuals are destined for some special bounty. Thus, chain letters, pyramid schemes, and lotteries are never in short supply. The simple truth is that the old adages "if something sounds too good to be true, it probably isn't" and "there's no such thing as a free lunch" are the best security advice you can have in the modern Internet-connected world.

Also, going back to my recommendations about e-mail file attachments, there are some viruses that masquerade as harmless, entertaining fluff. Simple applications that may show an animated cartoon or have a simple game may also be viruses. Most of these types of applications out there are completely harmless; however, they could just as easily be rifling through your hard drive while amusing you with a bit of topical humor.

It comes down to the fact that you simply can't know. Again, a virus-checker is the best defense, but not foolproof. If you want to look at an e-mail file attachment, simply waiting a day or two to see if any problems are discovered with the application isn't a bad idea. In addition, going to the web site of one of the anti-virus software vendors or groups should provide information about whether or not there is a security concern with a particular executable file.

It is often simply a matter of knowing which parts of your setup might pose problems.

Know thyself

The above phrase applies not only to you personally, but also to your computer setup. In terms of personal knowledge, look at what you plan to use your webcam for and think about what kind of security problems you might run into. Most webcam usage shouldn't be a very big risk, but again, something like a web site is a whole different ball game.

In terms of your computer setup, it is often simply a matter of knowing which parts of your setup might pose problems. An "always-on" connection can leave you open to hackers, and unencrypted web traffic can be "tapped." Usually all you have to do is figure out the risk and see if it warrants any kind of action. After all, not all risks are likely enough to worry about or are even preventable.

Never assume

This is perhaps the most often-made mistake when it comes to Internet security. People assume that their files are safe, that their anti-virus software will protect them, and that with all the traffic on the Web no one will bother with them. It keeps happening because, for the most part, it's true: the Internet is safe.

If you consider your safety in the real world, however, you would never consider walking down the street blindfolded with a few hundred dollars pinned to your shirt. After all, you wouldn't assume that you would reach your destination without running into something, or that you would reach your destination with the money still pinned to your shirt. Yet in a sense, that is exactly how some people behave on the Internet. They assume the anonymity that so many enjoy on the Net will protect them. What they don't realize is that same anonymity can protect hackers and scammers.

Looking out for No. 1

Don't automatically assume your ISP knows what it's doing in regard to security.

Sounds rather Machiavellian, doesn't it? My point, however, is that the only person truly interested or qualified to protect your computer is you. Your ISP and the providers of other on-line services usually do what they can to protect your interactions with them, but their focus is the bottom line, not the security of your information or your privacy. This ties in with what I said about not making assumptions about your security.

Don't automatically assume your ISP knows what it's doing in regard to security. If you simply have a connection, ask questions about whether encryption is available, whether it has a privacy policy, and what sort of protection it offers for things such as spam. If you have a company actually hosting a site for you, then you have even more questions to ask. Don't get a service based on the price alone, figuring they are all basically the same. Find out about the server platforms, the software used to post pages, firewalls, and the options for doing e-business on one of their servers. Not only will this information prove invaluable in evaluating different providers in terms of level of service, but it can also give you clues as to what security issues may be of particular concern.

Just as there can be a "wrong part of town," there can also be a wrong part of the Internet.

Finding the path through the wilderness

Walking or driving through a major city can be a lot like navigating the Web. You have your main thoroughfares that are usually the safest to travel. You may also run across alleys and dark back-streets that have a higher potential to be dangerous. Just as there can be a "wrong part of town," there can also be a wrong part of the Internet. Adult web sites, hackers, and other sites that are either on the edge of the law or over it can all be used by the unethical to breach your computer's security.

In ways both gross and subtle, these sites often lure the unwary. Once there, these sites may simply cause a nuisance (such as repeatedly opening new browser windows in an attempt to get you to part with your money), or they may try to get you to give up your credit-card information for goods or services of a questionable nature. Using security holes in your browser and operating system, they can sometimes even scan your hard drive or worse.

Whether you accidentally come across one of these sites or go to it on purpose, my advice is to be careful. Since these sites are often there one day and gone the next, don't count on being able to track down the owners if something happens to your PC or credit-card information. To sum up, the further you wander from the main drag, the more likely you are to run across trouble of some kind. This is not to say that the big commercial web sites are completely safe, but they have far more at risk in terms of public perception.

Forget your shoulder, look over your monitor!

The majority of people on the Internet will never have a problem with any real security issue, so why even worry? Well, just like seat belts, air bags, rubber gloves, and sneeze guards, some things are just good common sense. Anti-virus software is probably the biggest step you can take to ensure you don't have problems. Updating your operating system and applications with patches and fixes also can help close possible security holes.

All of that won't make a bit of difference, however, if you aren't careful in your on-line behavior as well. There's no real reason to be scared of using the Internet, but there are lots of reasons to be careful.

There's no real reason to be scared of using the Internet, but there are lots of reasons to be careful.

Appendices

In the Appendices that follow, I have listed a number of online resources—everything from listings of manufacturers to webcam community web sites. These resources will hopefully help to get you started on living your own "camlife." There is such a wealth of information and help available on the Internet these days that you will surely find something for whatever problem or question you have. I cannot list all of the resources available to you on the Net, as that would be a Herculean task, but I think what I have listed will provide ample assistance to you.

In addition, I have created a web site at www.mycamguide.com that contains some material from the book, as well as the links provided in these pages. So if you find a link that's no longer working, be sure to visit the web site for updated listings. You will also be able to contact me and share links you've found that will be added to my listings, and perhaps a few success stories from people who have used this book to set up their own webcams.

The listing of a company, product, or resource in these appendices is not meant to be an endorsement; rather I have simply endeavored to make the most complete listing of online webcam information for you.

Appendix 1: Webcam Software/ Hardware Vendors and Products

Webcam Hardware

3Com Corporation
3Com HomeConnect PC Digital WebCam;
http://www.3com.com/client/pcd/homeconnect/pcdigital/

ACS Innovations International Pte Ltd.
COMPRO PS39U Digital Video Camera;
http://www.acscompro.com/ps39/

ADI Corporation
PC Camera C350;
http://www.adi.com.tw/c350c.htm
PC Camera M350;
http://www.adi.com.tw/m350c.htm

Ariston Technologies
iSee-II USB Digital Color Camera (PC and Mac);
iSee-Pro USB Digital Color Camera (PC and Mac);
iSee-U2 2 USB Digital Camera Bundle (PC and Mac);
http://www.ariston.com

AVerMedia Technologies, Inc.
AVerMedia InterCam-Elite;
http://www.aver.com/products/intercam-elite.html

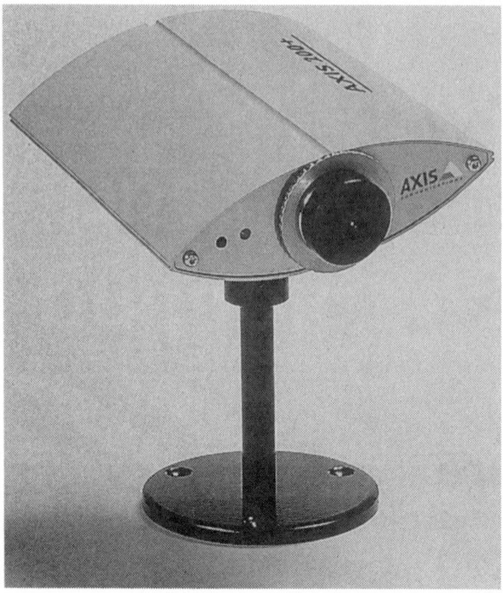

Figure A1-1. Photo of Axis 200+ courtesy Axis Communications.

Axis Communications
Axis 200+ Network Camera;
Axis 2100 Network Camera;
http://www.axis.com/products/camera_servers/

Creative Labs Inc.
VideoBlaster WebCam Go;
VideoBlaster WebCam 3;
VideoBlaster WebCam II;
http://www.americas.creative.com/video/

Figure A1-2. Photo of VideoBlaster WebCam Go courtesy of Creative Labs.

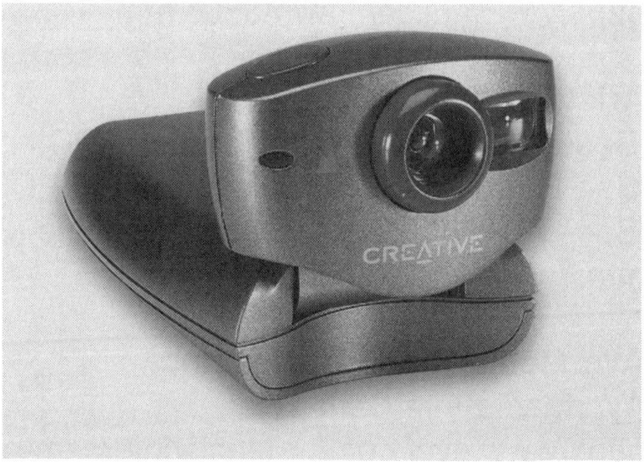

Eastman Kodak Company
Kodak DVC 325;
http://www.kodak.com/US/en/digital/cameras/dvc325/

Ezonics Corp.
EZcam USB Camera;
EZcam Pro Camera;
EZcam Plus Camera;
EZcam Dual Camera;
EZcam II USB Camera;
http://www.ezonics.com/pinfo.htm

Hitachi,Ltd.
Hitachi M2 3-in-1 Multimedia Recorder;
http://www.mpegcam.com/m2/

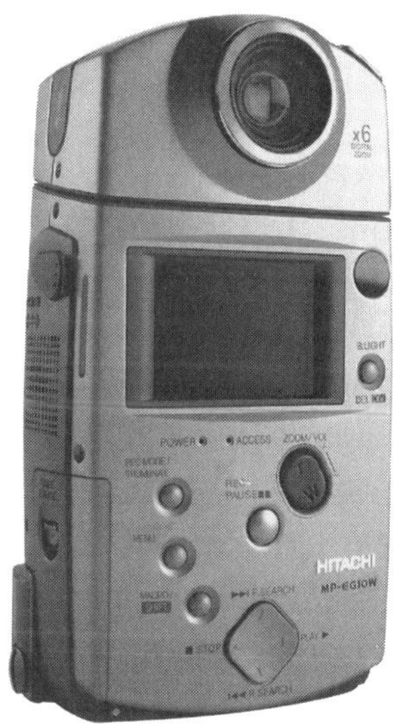

Figure A1-3. *Photo of Hitachi M2 3-in-1 Multimedia Recorder courtesy Hitachi, Ltd.*

IBM Corp.
IBM PC Camera;
IBM PC Camera Pro;
IBM PC Camera Pro Max;
http://www.xirlink.com/ibmpccamera/cameras.htm
http://www.pc.ibm.com/us/accessories/scanners.html

Intel Corp.
Intel PC Camera;
Intel PC Camera Pro;
http://www.intel.com/pccamera/

Kensington Technology Group
Kensington Super VGA;
Kensington VGA;
Kensington (Mac) VGA;
Kensington VideoCam;
http://www.kensington.com/products/pro_c1002.html

Logitech Intl.
QuickCam Pro;
QuickCam Web;
QuickCam VC;
QuickCam Express;
http://www.logitech.com

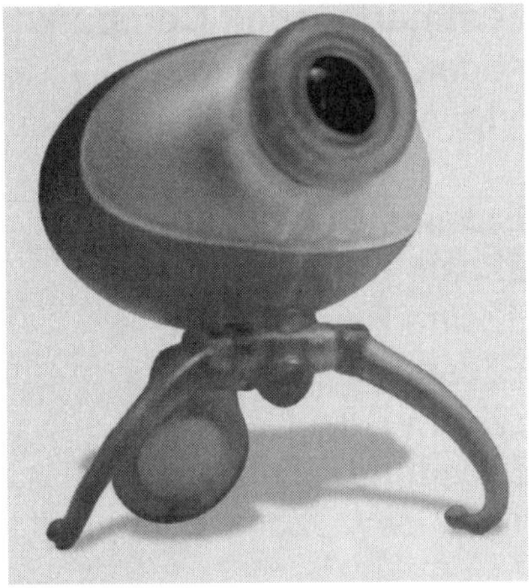

Figure A1-4. Photo courtesy PAR Technologies, Inc.

Microtek Lab, Inc.
EyeStar U2S Camera;
http://www.microtek.com

Panasonic
Panasonic GP-KR001;
http://www.panasonic.com/office/multimedia/multi.html

PAR Technologies, Inc.
The iRez Kritter USB Video Camera;
USB MicroCam Video Camera;
The iRez Kritter SV;
http://www.irez.com

Philips Electronics
Vesta Fun PC Camera;
Vesta Pro PC Camera;
Parallel Port PC Camera;
http://www.philips.com/productoverview/pcperipherals.html

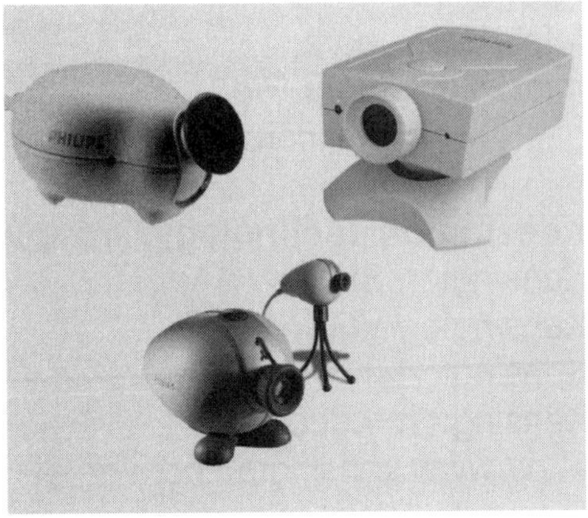

Figure A1-5. Photo courtesy Philips Electronics.

Sony Electronics Inc.
Sony CCD-PC1;
Sony CCD-Z5;
Sony CCD-Z7;
http://www.ita.sel.sony.com/products/imaging/

StarDot Technologies
WinCam.Live Camera;
SecureCam Pro;
http://www.wincam.com

Utobia Corporation
Utobia USB Digital Video Camera;
http://www.utobia.com/camera.htm

Vista Imaging Inc.
ViCam LPT Digital Camera;
ViCam USB Digital Camera;
http://www.vistaimaging.com/products/retailproducts.htm

Zoom Telephonics Inc.
ZoomCam USB (Mac);
ZoomCam USB (PC);
ZoomCam PPC;
ZoomCam ISA;
http://www.zoom.com/video.html

Webcam Software

Cam Shot

Company: BroadGun Software Ltd.
Description: CamShot is a Windows 95/98/NT web server that serves up web pages containing time-stamped images captured from a video camera.
Price: $25.
http://www.broadgun.com/camshot/

Capture Studio

Company: Pegasus Imaging Corp.
Description: Capture Studio is an image- and video-capture program. With it you can capture still-frame images and

full-motion video. You can use CaptureStudio to set up unattended periodic events to capture and save frames locally, send by e-mail, or post to FTP sites for use with webcam Internet sites.
Price: $19.
http://jpg.com/webcam/

ChillCam
Company: Viziva Software.
Description: ChillCam saves images from any Video for Windows device, saves the jpeg to your hard drive, and/or FTPs it to a web server. You can have any number of schedules, and can send to any number of different FTP servers. In addition, you can overlay any number of logos or watermarks and independently set their transparency.
Price:$25.
http://www.thegeek.com/chillcam/

Easy Snap
Company: O'Reilly Solutions.
Description: EasySnap is webcam software that works with Windows 95, 98, or NT, and the Snappy video-capture device from Play, Inc. In combination with the Snappy hardware, it can capture images from any video source such as a video camera, digital camera, or even video games. It can then save them and upload them to a web page. It can also be used to schedule image grabs and will number the sequential files for easy indexing.
Price: $25.
http://www.easysnap.com

Gotcha!
Company: Prescient Systems.
Description: Primarily marketed as security software, Gotcha! can turn your webcam into a motion detection-based video surveillance tool. It captures images only when triggered

(primarily through changes in light and dark areas within the picture frame), providing time-stamped video surveillance files of events as they occur without recording inactivity between events. Gotcha! provides many video-surveillance features, including an e-mail notification video-surveillance feature that lets you know what is happening at home. You can also use the FTP image upload video-surveillance feature to create your own web video surveillance site.
Price: $69.95.
http://www.gotchanow.com/

GTS Interactive Media Server Release 3.1
Company: Graham Technology Solutions.
Description: IMS can be used in a wide variety of visual-monitoring, remote-collaboration, and interactive-broadcast services such as plant security, process validation, Internet seminars, corporate announcements, and traffic management. Able to leverage existing closed-circuit TV systems and the new video-server appliances, IMS enables camera viewing securely over the Internet.Depending on the type of video server or CCTV system, the software allows authorized users to control camera operations—such as pan, tilt, and zoom—over the Internet. The system can also be used to set up videoconferencing solutions.
Price: License and service-based pricing.
http://www.graham.com

How'sMy
Company: HowsMy.com, Inc.
Description: HowsMy.com is an Internet-based subscription service that allows you to send and view live video of your child, pet, business, or anything else via the Internet.
Using a video camera or webcam, a computer, and an Internet connection, you can view your own live video from anywhere in the world. With HowsMy.com's security, you can ensure that only you and others you authorize have access to the

video feed you have designated.
Price: Subscriptions costing anywhere from $79.95 a month to nothing, depending on the number of cameras involved, number of users, and so on.
http://www.howsmy.com

iVista
Company: InetCam.
Description: Works on any PC with Windows 95, 98, or NT operating systems, and it creates a direct point-to-point connection for live video transmission. InetCam video can be viewed on any Internet-connected device that supports Java and requires no special downloads or viewers. The video stream can either be attached to an existing web site or located on the Web by means of a dynamic host name provided by Inetcam at no charge. iVista works with almost any existing camera format, and up to four video streams can be broadcast simultaneously by the operator.
Price: Free 10-day trial version or $30 for full version. Company also sells combined hardware/software solutions.
http://www.inetcam.com

ISpy
Company: Surveyor Corp.
Description: ISpy webcam software grabs video images, saves them as jpeg, and sends the image automatically to your web site. ISpy runs under Windows95, Windows98, and WindowsNT 4.0., and works with any Video for Windows compatible video-capture card, camera, or frame-grabber. You can schedule when the images are captured and sent. In addition, ISpy now supports live video with continuous uploads or image-loops using a simple Java script. Also included with the purchase of ISpy is Surveyor Corp.'s Webcam32 software (through on-line download.).
Price: $25.
http://www.surveyor.com

KABcam
Company: KAB Software
Description: KABcam takes the images from any Windows compatible camera or video device and uploads it to your Internet site. Transfer can be scheduled using the software, and includes time stamping. Latest version includes a Java applet that allows the client side to pull a new image from the camera, rather than waiting for an update.
Price: $10.05.
http:// www.kabsoftware.com/

NetSnap
Company: Logan Industries, Inc.
Description: NetSnap publishes images from your Video for Windows camera or Snappy device to the Internet. Included in the package is a built-in HTTP server and support for streaming video. The HTTP server can be password protected to secure cam images, and the included scheduler can not only be used to determine when images are captured and transferred, but can connect your dial-up connection automatically as needed.
Price: $49 (on-line purchase price) for single-image stream. Four-image streams cost $89, and 16 streams cost $197. Other streaming amounts available.
http://www.netsnap.com

Oculus (Mac)
Company: Poubelle Software and International Web, Inc.
Description: Oculus is a Macintosh program that allows you to capture video frames, caption them in a variety of ways, and deliver them via the World Wide Web. It has options to transmit an image to a remote FTP site or save a file locally, if you have a web server running on your video-capture machine. Oculus can capture images into both Photo and Movie formats. This software features a flexible captioning architecture that allows for text and picture captions to be

randomly or sequentially applied. In addition, it uses features of QuickTime 3.0 to apply filters to picture captions or to the final image itself.
Price: $20 single-user, $500 for a complete site license.
http://intlweb.com

SiteCam (Mac)
Company: NuSpectra Multimedia, Inc.
Description: SiteCam 5.0 is a Macintosh webcam application for distributing live images and time-lapse movies via the Internet. It includes a built-in HTTP server that is performance-tuned to serve real-time streaming video to a browser (Netscape or Internet Explorer) without the need for third-party viewing software. It can automatically select the best streaming format for the viewer's browser and stream images to fit the viewer's available bandwidth. In addition to streaming video, SiteCam can also be used to provide still images to a remote server using FTP.
Price: $149.
http://www.nuspectra.com/sitecam/index.htm

SnapCAP
Company: ARTA Media Group.
Description: SnapCap is a complete image-capture application, working with Video for Windows devices and Snappy. Included in the package are picture scheduling, incremental archiving, a continuous capture mode, and flexible image sizing. Also included are built-in HTTP and FTP servers, thumbnail generation and atomic clock synchronization.
Price: Free.
http://www.snapcap.com

SpyCam
Company: Bill Oatman.
Description: SpyCam captures images from Video for

Windows compliant devices and uploads them to any web server via FTP. Updates can be scheduled, and different update rates can be set for guests, if you have a subscriber-based webcam. You can save an unlimited number of archive images, and settings are available to change image quality settings, size, and file compression.
Price: $15.
http://www.getspycam.cjb.net/

TrueLook
Company: Perceptual Robotics, Inc.
Description: PRI provides a suite of video-over-the-Internet applications for everything from asset management to adding a live video component to e-commerce business sites. With TrueLook/Live, TrueLook/Enterprise, Look and Buy, and Look and Bid, PRI's focus is solidly on helping businesses make video from webcams or other sources a productive and profitable part of their operations.
Price: License and service-based pricing.
http://www.perceptualrobotics.com/

TrueTech WebCam
Company: TrueTech B.V.
Description: There are three editions of this software available: Personal, Professional, and Commercial. The Personal Edition is completely free, but is limited to 10 simultaneous connections and does not have the ability to stream audio. Upon registering the free version, however, you can get a free web page to post to on the company's cam-community site, Camarades. The Professional and Commercial editions are aimed specifically at those who wish to use their webcams and streaming audio/video as part of a business web site. These versions can be used as part of a relay network for distributing the bandwidth load for streaming-video broadcasts to many users. The Professional Edition is licensed for one machine and an unlimited number of relay

servers, while the Commercial Edition is licensed for the same number of machines, but includes an e-commerce module for setting up subscription services on video streams.
Price: Personal Edition - Free; Professional Edition - $450; Commercial Edition - $6,000.
http://www.truetech.com

Video Control 98
Company: STEFRA Corporation.
Description: Video Control™ is a Windows utility that can quickly capture and edit video, still images, and sound from digital cameras and other image sources. Once you have captured and edited video clips, audio, or images, Video Control™ can also perform videoconferencing.
Price:$49.
http://www.stefra.com

Webcam32
Company: Surveyor Corp.
Description: Webcam32 is probably the most well-known and widely used webcam software today. Specifically designed for use with webcams, Webcam32 features on-screen wizards to help you get up and running quickly. The software can be used to schedule image captures and transfers, add captions and image overlays, as well as time stamps to your images. Webcam32 can also be used to broadcast streaming video or just an audio stream with your still image. This package also allows you to set up on-line chats as part of a webcam site, and automatically generates a complete HTML-coded page if you wish. It can be used with any Video for Windows camera or capture device.
Price: $25 or a free one-hour downloadable trial version.
http://www.surveyor.com

webcamnow

Company: webcamnow.com Inc.

Description: webcamnow is a combination of image-capturing software and a webcam-hosting service. Included with the hosting service is the ability to have on-line voice and text chats with viewers. Accounts begin with a free version that allows video streams of six frames per minute. A "Pro" version is available that provides 30 frames per second, thumbnail previews, and a priority listing in its community listings. The company will also soon provide a business-oriented broadcast capability.

Price: Basic version is free; webcamnow Pro is $9.95 a month.

http://www.webcamnow.com

Appendix 2: Resource Web Sites

Webcam Tutorials, Resources, and Advice

4Anything.Com
Description: This site brings together a number of outside resources in one easy-to-find page. A number of the links were a little dated, but overall, this is still a good page.
http://4webcams.4anything.com/

About.com
Description: This is a truly excellent site to find out almost anything about digital video cameras. Made up of information produced "in-house," there are not a lot of worries about finding a promising link and then finding out it is out of date. The depth and breadth of information to be found here is staggering, with everything from videoconferencing to webcams, whiteboards, and cam-parody sites. This is one of the places I recommend you come to first when you have a problem or question.
http://netconference.about.com/internet/netconference/mlibrary.html

StreamingCam.com
Description: This site is interesting in that it concentrates completely on helping with streaming-video questions. It includes reviews of products and books, lists of streaming-video cam websites, and news about webcam technology

and companies. It also includes an on-line store that sells hardware and software.
http://www.streamingcam.com

Top Dog.Com - Tutorials
Description: The main page of this site features a directory listing of webcam sites, but don't be fooled, it has a first class tutorial area as well. While it is as yet incomplete, the items it does provide help on are well written and thought out. A bonus is the inclusion of templates with actual code to make your own webcam page.
http://www.topdogcams.com/tutorials.html

Webcam Community How To
Description: This is another how-to section from a more general cam directory site. It does have some useful hints on how to get started, but I found the pink and purple text a little hard to read for very long!
http://webcamcommunity.com/howto.htm

Webcam Developers
Description: This site has the whole soup-to-nuts guide to webcams. Everything from hardware and software lists to accessories and sample scripts and templates. Written in a clear, concise manner, this site can be a big help to just about any level of webcam user. From the company that makes the WebCam32 and ISpy software packages, there's obviously a slight editorial slant, but much less than you might expect.
http://developers.webcamworld.com/

The Webcam Resource
Description: Again, this is part of an overall cam directory site, so the tutorial section is actually a list of links to tutorials, rather than an actual tutorial. The site listing is quite

extensive, however, and the individual links are ranked by visitors to the page. If a search-engine query leaves you more confused than when you started, this site is a good remedy.
http://www.camcity.com/webcamresource/Tutorials/

ZDTV Netcam Network
Description: As part of the Ziff-Davis ZDTV enterprise, this site attempts to involve cam users with the shows on ZDTV. As part of the site, the "Tips and Tricks" section contains lots of guides on how to do some unusual things with your cam, including stop-motion animation and other special effects.
http://www.zdnet.com/zdtv/interact/tipstricks/

Cam Tools and Extras

AmberCam
Description: The *AmberCam* applet magnifies your webcam image 200 percent. This reduces bandwidth or allows you to update the image more often. For example, blow up your 160 x 120 pixels image to 320 x 240. The applet uses a simple interpolation method that makes the image a bit less sharp.
http://www.castleamber.com/ambercam.html

Anfy Java
Description: This collection of Java applets allows you to add special effects to your webcam images and site—everything from floating cubes to special image "morphs." Definitely not necessary, but certainly fun.
http://www.anfyteam.com/ajdown1.html

CamSwitch
Description: Automatically replaces a live image with a

"please-stand-by" image of your choice after a set time if no new live image is sent. This can keep your web site visitors from thinking the cam is off when you experience delays.
http://fathom.org/camswitch/

ICQ
Description: Probably the premier Instant Messaging tool among cam users in the United States, if not the world. While the primary use of Instant Messaging (or simply "IM") is to have on-line text chat, the program can also be used to indicate a person's on-line status. Many cam operators use it solely for this purpose, allowing potential visitors to see immediately if they (and supposedly their webcam) are on-line. Don't forget about this tool's IM features, though, as it is simply one of the best Instant Messaging applications out there. Did I mention it was free?
http://www.icq.com

OIS Webcam Enhancer
Description: WebCam Enhancer consists of Image Refresher and the Status Indicator. Image Refresher makes the image itself refresh, without the entire web page reloading, and without a Java applet. The image will update live without the user's screen flashing. The Status Indicator is an on-line/off-line indicator for your webcam. You can choose from three different sets of images to use, whichever one best suits the look of your page. This is more of a service than a product, and thus charges a monthly fee, which depends on what type of cam site you are running.
http://webcam.ois.tj/join/

MSN Messenger
Description: Since Microsoft closed down its NetMeeting conferencing servers in favor of those provided by independent third parties, MSN Messenger has become *the* tool for those wishing to meet up on-line, either using MS

Messenger, NetMeeting, or webcam sites. Once again, this is a primarily IM-focused application used by many as an on-line status indicator.
http://messenger.msn.com/

Yahoo! Messenger
Descriptions: From the people who brought you one of the most popular search engines comes another IM application. While not quite as full-featured as ICQ, I have seen some webcam sites use the on-line status indicator for the same purpose. And, once again, it's free!
http://messenger.yahoo.com/

Web Directories

A1 Webcams
http://a1webcams.monsterserve.com/

AllCam
http://www.allcam.com/

Ambit Live Web Cams
http://www.ambitweb.com/cams/cams.html

Camarades.com
http://www.camarades.com

Camcities
http://www.camcities.com/

CamCity
http://www.camcity.com

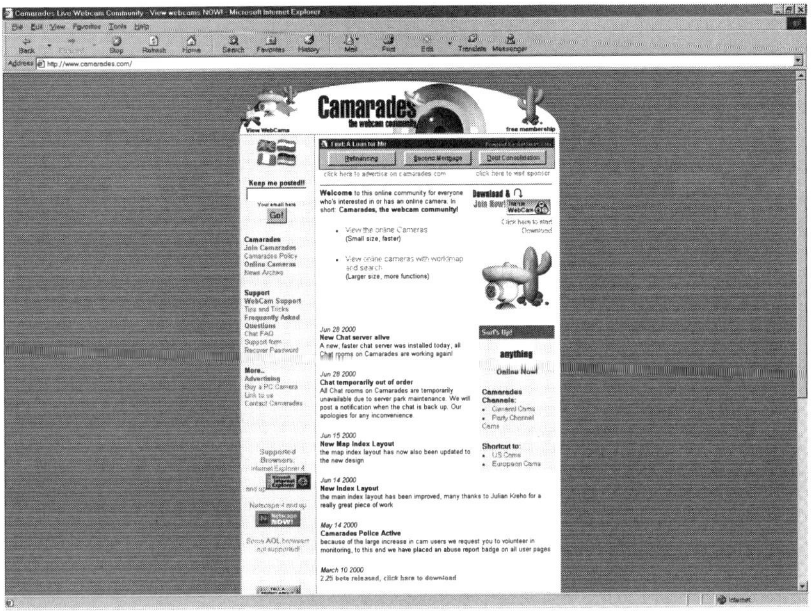

Figure A2-1. Screen shot of Camarades.com.

Cam.Com
http://www.cam.com/cam/

CamTastic
http://www.camtastic.com

Camville
http://www.camville.com

ConnecTv
http://connectv.2nd.net

Earth Cam
http://www.earthcam.com

Extreme Web Cams
http://extreme.homepage.com/cam.html

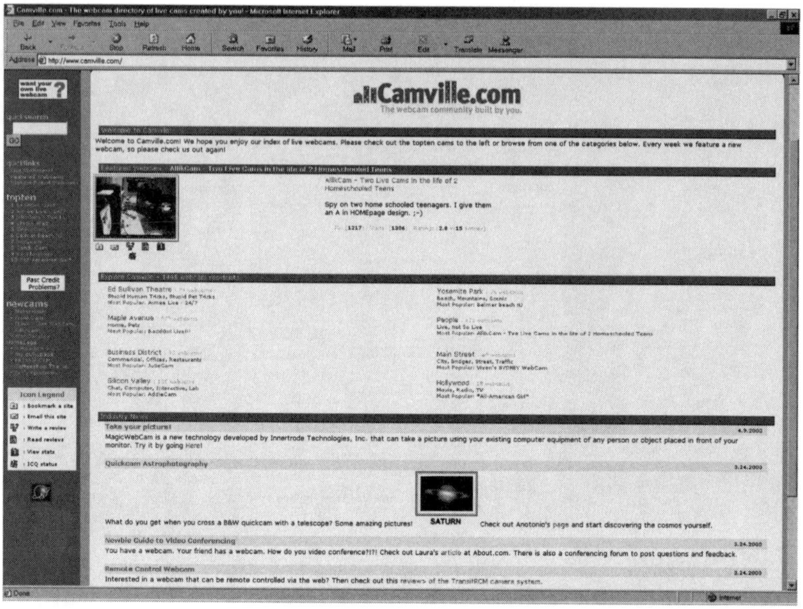

Figure A2-2. Screen shot of Camville.com

I Ontheworld
http://www.iontheworld.com

Live Weather/Tower Cams
http://www.weatherimages.org/weathercams.html

Online Camera
http://www.onlinecamera.com

Peeping Tom
http://www.coolbase.com/peepingtom/index.htm

SpotLife
http://www.spotlife.com

Szym.com
http://www.szym.com/cameras/

TopDogCams
http://www.topdogcams.com

Video Dog
http://www.video-dog.com/main_e.htm

Webcam32 Gallery
http://www.webcam64.com/gallery/page1.htm

Webcam Center
http://aaabis.simplenet.com/webcameras/index.html

Webcamania
http://members.brabant.chello.nl/~s.storms/webcamania/

WebCam Resource
http://www.camcity.com/webcamresource/index.shtml

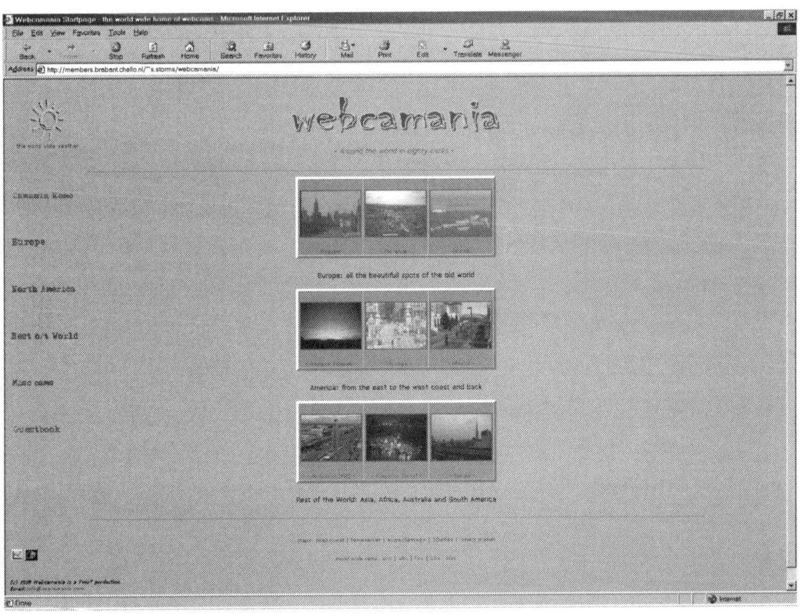

Figure A2-3. Screen shot of Webcamania.

Webcam Search
http://www.webcamsearch.com

WebCam World
http://www.webcamworld.com

Usenet Newsgroups

Newsgroups can be a priceless resource for webcam enthusiasts. Accessible through newsreaders and e-mail programs such as Outlook Express, newsgroups are simply message boards. Please note that not all newsgroups are subscribed to by all Internet Service Providers. If you see a newsgroup here that you can't access, please contact your ISP's support group.

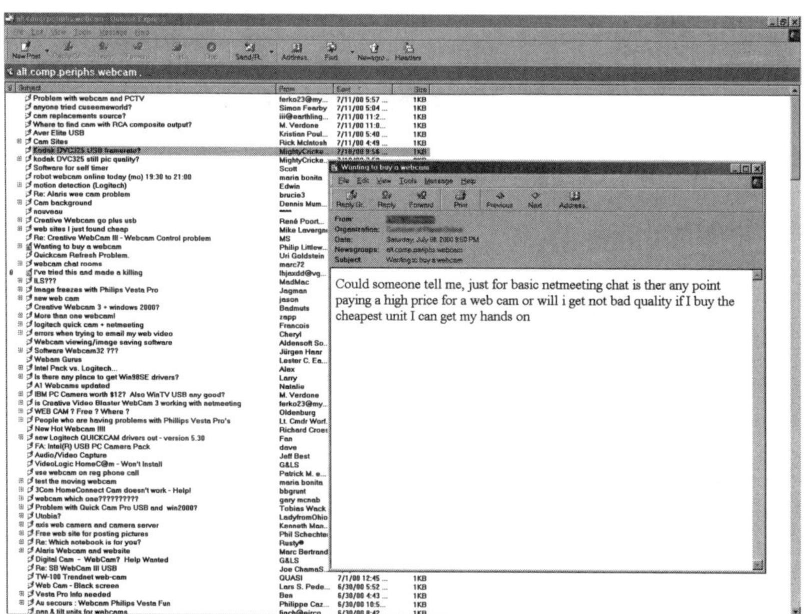

Figure A2-4. Screen shot of the newsgroup alt.comp.periphs.

Newsgroups exist by the request of users, and may exist for specific technologies, products, or services. For example, in the list that follows, there is a newsgroup for users of the VideoBlaster Webcams from Creative Labs. These vendor-specific newsgroups are often monitored by technical-support specialists at the company, and if no one else has a solution to your question or problem, they might.

While newsgroups are a great place to share and find solutions to problems, be aware that those companies and individuals who use e-mail to mass-market products and services (otherwise known as spammers) often pick out e-mail address from newsgroups to get more names to mail to. If you are worried about spam, and who isn't, then set up your newsreader to show your return e-mail address as something like: youremail@nomail.youremailaddress.com. This will stop most attempts to harvest your address for spam, but still allows those who wish to contact you directly to do so. Simply put a note at the end of your message about what part of the listed e-mail address to remove to correctly address e-mail to you. Below find a list of the seven most common webcam-related newsgroups.

1) alt.comp.periphs.dcameras
2) alt.comp.periphs.webcams
3) comp.dcom.videoconf
4) creative.products.video_blaster.webcam
5) microsoft.public.internet.netmeeting
6) rec.photo.digital
7) rec.video.desktop

Appendix 3: Favorite Webcam Sites

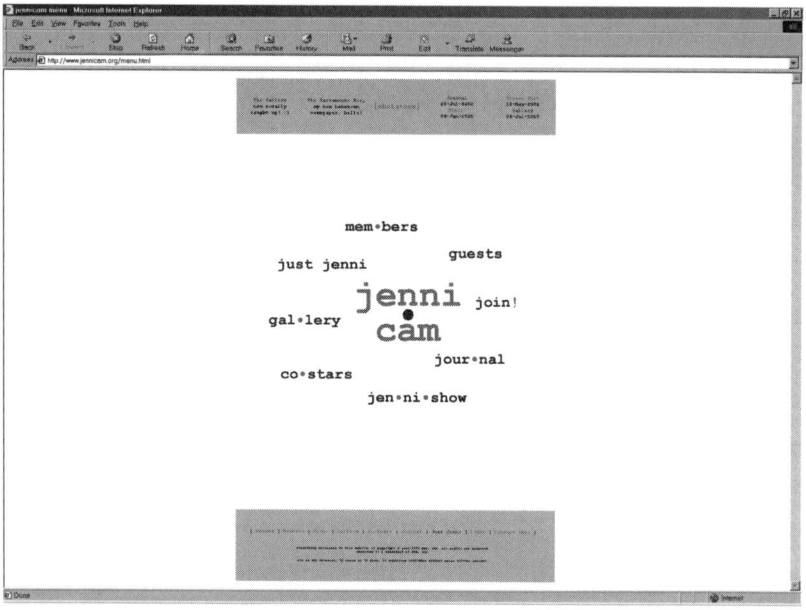

Figure A3-1. Screen shot of JenniCam.com.

www.JenniCam.com

This one, as I said earlier, is what pretty much started it all in terms of personal webcams. Jennifer Ringley has had countless print interviews published, been on talk and news shows, and has even made an appearance in a small movie. Hers is also, from a design point of view, one of the most polished webcam sites out there. She does most of the work herself (as she is a freelance web designer) and is constantly updating her site.

Things to take notice of are how clean her site design is, and how the cam is only a part of the site. In the spirit of having her life be an open book, she has posted a journal, biographical information, pictures of her pets, and all sorts of little digital knick-knacks. Her page design also is uncluttered and easy to navigate, meaning you always know where you are and it is usually easy to find what you are looking for.

So not only was this site a first, but it also continues to captivate and entertain people through a combination of relatively uncensored access to a real person's life and a solid design. If you're considering making your own webcam site with either a personal or professional slant, take a look here and see how it is done right.

www.anacam.com
This site is one of my personal favorites, not because of any single thing it does, but just for the obvious creativity and force of personality of its subject, Ana Voog. Singer, performance artist, and just plain unusual person, Ana's anacam.com is like a grab bag at a thrift store. You never know what you'll find, but it always seems worth it.

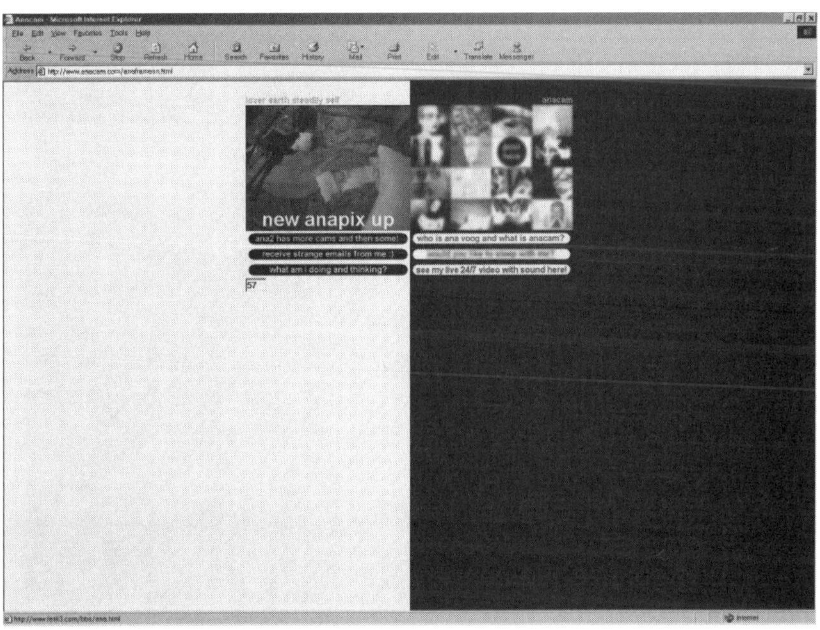

Figure A3-2. Screen shot of anacam.com.

AnaCam.com is much more about art than reality. Not that Ana doesn't talk about the details of her life in her on-line journals and e-mail lists, but the cam itself is usually used by her to express artistic ideas. Rarely will you find a simple unaltered view of her on the cam, instead you'll see her image through strange filters and surrounded by props such as one of her collection of mannequins. On her paid membership site www.ana2.com, she performs live performance art pieces. While certainly not everyone's cup of tea, anacam.com is the perfect example of how artists and just generally creative individuals can take technology and turn it into something entirely new and, at the same time, quite human.

http://www.ultimatetaxi.com

What, did you think all webcam sites were run by women? While it may seem that way sometimes—and the majority certainly may very well be—there are webcam sites run by men as well. In fact, while a webcam run by a guy is unusual, the most unusual thing about this cam site is the fact that it displays images from a taxicab.

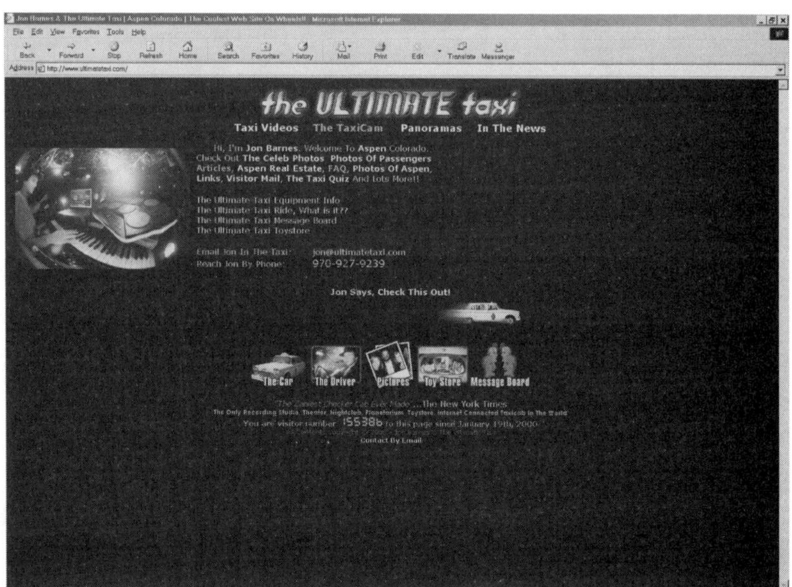

Figure A3-3. Screen shot of ultimatetaxi.com

Like the site, the cab itself is a one-of-a-kind creation. Based in Aspen, Colorado, Jon Barnes doesn't provide a normal taxi service, but instead what he calls, "a 30-minute musical, magical adventure around Aspen." During his cab rides, images of the passengers are broadcast, via a wireless Internet connection, to a web server. Jon also takes and sends e-mails, and occasionally even stops by his own chat room, though he says he pulls over and stops to do this. This site is a perfect (though unusual) example of how a webcam can be used to enhance an existing business.

www.nerdman.com

This is another site featuring a man, this time a self-professed nerd with a small obsession with webcams. Using a total of 18 cams on his site, we definitely think he qualifies. He has six digital video cameras at his house and 12 at his place of work; he definitely spends most of his time in front of a cam of some sort. And before you think that he must have no life, he is indeed married and they recently had a child.

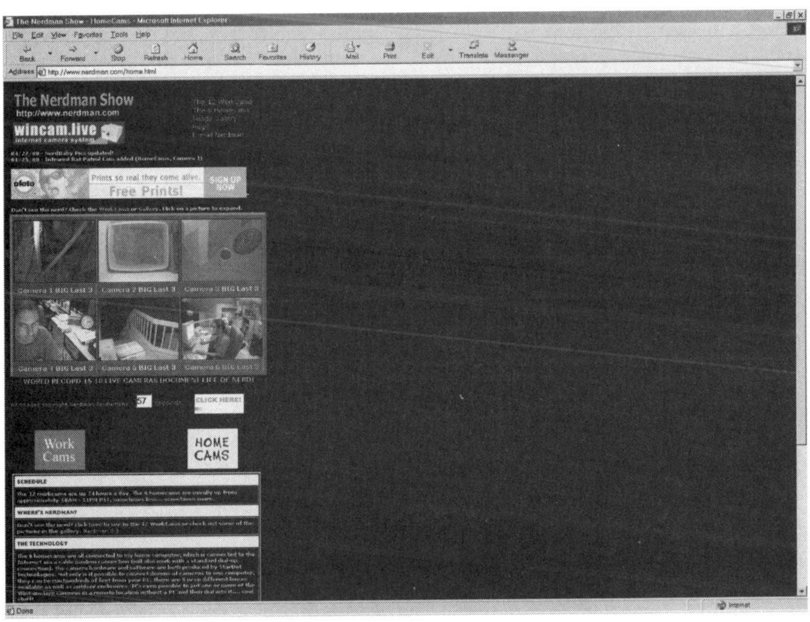

Figure A3-4. Screen shot of nerdman.com.

When you visit this site, you quickly get the idea that he has so many cameras because of a desire to see if he could do it, rather than a blatant need for attention. He even has one of the cams in his attic to see if he can spot a bat he believes to be up there—certainly an interesting use of the technology.

He has been running his site since 1996 and now uses a cable-modem connection and a single computer to run all six of his home webcams. When he first started, however, he was using only a 33.6Kbps dial-up connection. The webcams at his office update 24 hours a day, and the home cams update only when he is awake. This site is also nice because, unlike some, he tells you exactly how he does it all and provides links to the companies he used to set up his site.

www.artiscam.com
This is another unusual webcam site featuring two cameras that you can control through a series of buttons on the bottom of the screen. What could be so interesting

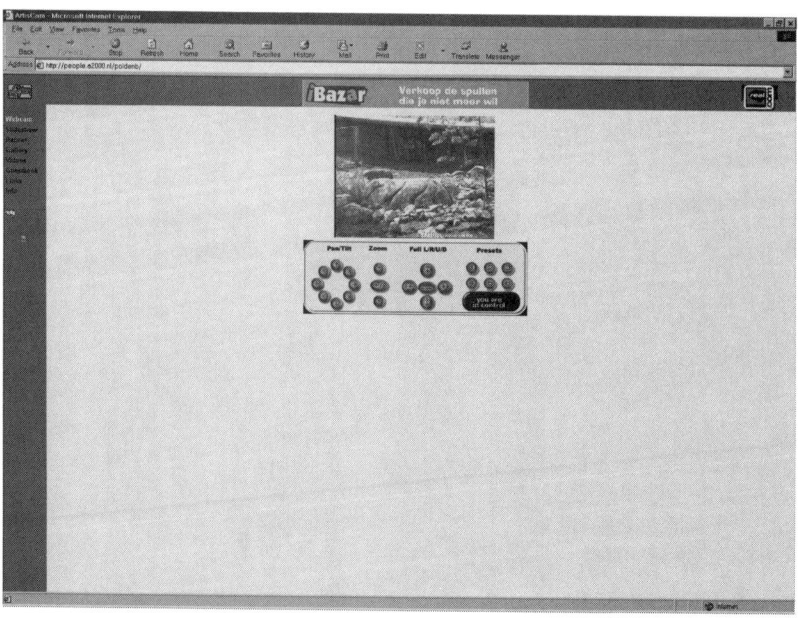

Figure A3-5. Screen shot of artiscam.com.

that you would want to be able to pan the camera around to any number of angles, and even zoom in and out? How about a zoo? The site was actually put up by somebody living in a building overlooking the Amsterdam zoo called Artis. While the Internet in general (and cam sites specifically) has a seedy side, there seems to be almost as many sites featuring pets and wild animals as there are featuring more adult entertainments.

This site also demonstrates that webcams are hardly a distinctly American phenomenon. It is really something when you think that you can get on-line and look at a zebra in an Amsterdam zoo one minute and a shot of Mt. Fuji in Japan the next. Another thing to notice with this site is that the camera controls in this case do provide a useful service—namely, being able to follow animals around their enclosures. Oftentimes you may see cam controls like these on a site, but they really don't add very much to your experience. This is an entertaining exception.

http://www.pbs.org/wgbh/thisoldhouse/home.html
Webcam sites are not just popular with individuals and business trying to get customers, but TV shows even have them now. While normal broadcasting is still done over the airwaves, the PBS how-to show *This Old House* uses webcams in its renovation projects to give viewers a behind-the-scene view into the day-to-day activities at whatever property being working on.

The show has three cams set up in the current project: one giving a view from the outside, one in the basement, and one on the second floor. This site also has interactive polls about different facets of the project. The polls don't actually decide how the project goes along, but it and the webcam images provide an interesting facet to the program beyond its half-hour installments.

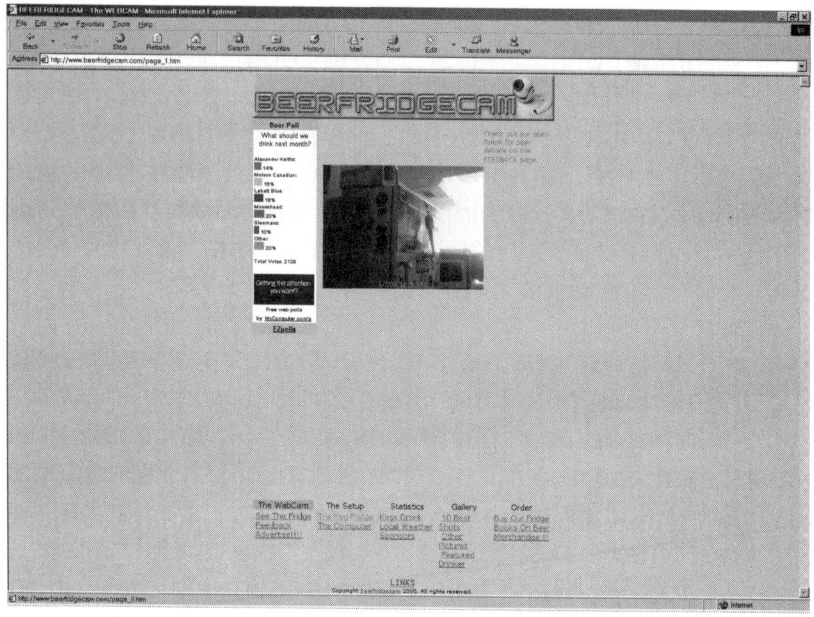

Figure A3-6. Screen shot of beerfridge.com.

http://www.beerfridgecam.com/

This is certainly one of the stranger sites I've happened upon—and certainly one of the least exciting. It does, however, have a certain something to it—namely, large amounts of beer. The owners of this site obviously like their beer quite a bit, so much so that they have converted an old refrigerator into a keg, complete with a tap on the front door that dispenses their beer. Now this would seem pretty extreme behavior under any conditions, but then what good is doing something noteworthy if no one notices?

That conundrum seems to have driven these fellows (they had to be guys, didn't they?) to create this site. Featuring a constantly updating image of their creation (it really has to be one of the most pointless uses of this technology to date), the site goes into great detail about how to build one of your own beer-fridges, information on what beer they are currently drinking, on-line polls on what they should drink next month, and other critical minutiae such as what beers they drunk in the past.

You may have noticed that all the webcam sites I've mentioned here are examples of different types of sites; so what category does this one come into? It is one of a number of sites out there that seem to have no discernible purpose but to showcase a weird or unusual feat of some sort. It is certainly strangely fascinating seeing what someone will devote a webcam to next. This site is of particular interest because of the odd subject matter mixed with a well-thought out and developed web site. If you're going to brag to the world about your latest invention or creation, you might as well do it right!

http://www.sloppyjoes.com
This is a more typical example of a business-related webcam site. Sloppy Joe's is a well-known bar in Key West, Florida. The original Sloppy Joe's was a favorite of Ernest Hemingway, and while this is not the original, it is a favorite of both locals and tourists. The bar does a lot of heavy-duty marketing, including T-shirts and other memorabilia, and is quite well known already, so why bother with a webcam?

Well, every little bit helps, of course, and the site comes up on many search engines when people are researching places to go in Key West. Having a webcam provides that little extra edge in getting people to the site—and hopefully to the bar when they visit.

The cam doesn't show much, just still images of the interior of the bar and occasionally a live musical performance (no streaming or sound, though). But the site does include the history of the bar and lots of information about upcoming events. The site would be interesting on its own, but the webcam probably helps bring in a lot of traffic.

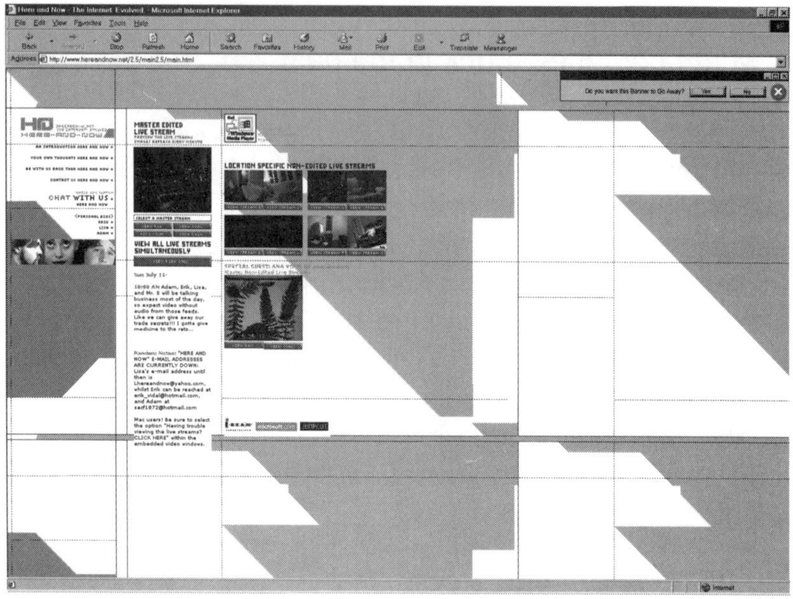

Figure A3-7. Screen shot of hereandnow.net

www.hereandnow.net

Most of the sites I've talked about focus on the traditional still image-based webcam site. Hereandnow.net is more than a little different. Think streaming video and sound in an MTV *Real World*-type environment. It's a lot less entertaining than professionally cut and edited "Reality TV," but it is certainly more real. That's both its strength and its weakness.

The site, while only occasionally entertaining, seems to certainly fascinate more than a few people out there. Live and unedited access to someone else's life, no matter how mundane, is still a rarity. With more webcam sites looking at starting up streaming video (though it is only the large ones that can afford it for now), it is interesting to stop by this site and see how it started.

Also, think about how long it takes you to download most webcam images, and then think about the bandwidth needed to push streaming video to thousands of viewers at once.

Some day we'll get used to people living their lives like Jim Carey's character in the movie *The Truman Show*. When that day comes, however, the most interesting thing—as it is now—might be how the whole thing is produced to begin with.

Glossary

In Chapter 3, I defined some webcam-oriented terms, but this Glossary lists general PC terms. Though by no means exhaustive, this Glossary does contain terms I believe to be the most helpful in understanding the general workings of PCs.

ADAPTER CARD: Used to describe a circuit board that plugs into your PC's motherboard. An example would be a graphics card or network card. Adapter cards add some capability beyond those provided by your basic system, and thus are sometimes called "add-in cards." On PCs there are different types, mainly ISA on older computers and PCI on newer ones. Graphics cards also come in PCI versions and AGP. They have differently configured slots to ensure that they are not plugged into the wrong kind of connection.

ADDRESS: Location of a web site or file on the Internet. It can either be a raw IP address such as 255.255.255.255, or a URL address such as www.website.com. It is also used to refer to File Transfer Protocol addresses, such as ftp.website.com

AGP: Accelerated Graphics Port. Type of video adapter card that provides better speed and memory handling over traditional PCI cards. Almost all new computers ship with AGP capabilities, while older computers usually cannot be upgraded to AGP because they lack the proper architecture.

ARCHIVE: Collection of older files. With webcam sites, an archive is sometimes available for visitors to view previously captured images.

BACK UP: Copy of files made to either free disk space or to ensure against possible data loss from a PC crash. In the past, this usually meant saving to a special tape drive, though now you can save to rewritable CDs, recordable CDs, Zip disks, and other high-capacity removable media. It is generally a good idea to back up your files on a regular schedule, usually anywhere from a week to a month, depending on how much you use your computer. If you have a web site, make sure to back up all your files in case of server problems or hacking.

BOOKMARK: Term primarily used with Netscape Navigator, but also used generically to describe the saving of a web address in your browser software for easier future access. In Internet Explorer this is called Favorites, and in AOL it is called My Favorite Places.

BROWSER (WEB BROWSER): Software program that enables you to see and hear content on the World Wide Web by rendering a coded HyperText Markup Language page. Browsers, for the most part, follow recognized industry standards, though every company may have different "interpretations" of those standards. This results in slight differences between Internet Explorer and Netscape Navigator, for instance.

CACHE: Used to refer to any use of memory that enhances performance. In regards to browsers, a copy of the web pages you've already visited remains on your hard disk in a web cache. When accessing the page again, unchanged items are pulled from your cache, rather than reloading them across your Internet connection. Especially with 56Kbps dial-up connections, this can increase your subjective speed of browsing. In regard to computer hardware, a cache is the memory used to store data between the processor chip and the main system board. The larger the cache, the better able the computer will be to multi-task.

CD-ROM: Compact Disk - Read Only Memory. A disk drive that can read data stored on standard compact disks. It is the current standard that 90 percent of all programs ship on.

It is starting to be replaced somewhat by DVD drives, but all DVD drives will read CDs as well. CD-ROMs, however, cannot read DVD disks for the most part.

CD-R: Compact Disk - Recordable. A drive that can be used to write information to a special CD-R disk. This process can only be used one time, whereupon the data on the CD-R disk becomes permanent like a normal CD. When making a CD with a CD-R drive, the slang term is "burning" the disk. A CD-R drive can also read CDs like a normal CD-ROM drive.

CD-RW: Compact Disk - Re-Writeable. A drive that differs from a CD-R drive in that the special CDs it uses can be written and erased many times, much like a standard floppy disk. A CD-RW drive can also be used like a standard CD-ROM drive.

CPU: Central Processing Unit. While the CPU is technically part of the main processor in your computer, the term is commonly used to collectively refer to the case of your computer and its contents. Almost all calculations are either processed by or routed elsewhere by the CPU. The faster the CPU, the faster the computer.

CRASH: Used to describe the situation when your computer doesn't respond to your input anymore. Often accompanied with an error message, it can be caused by incorrectly configured hardware or problems associated with either applications or your operating system. Often the only remedy is to power down your computer and turn it back on or to press the "Reset" button if your computer is so equipped. Try to make sure that you note as much as possible the exact circumstances of the crash to make diagnosing the problem—either by yourself or with the help of technical support personnel—as easy as possible.

CRC: Cyclic Redundancy Check. Standard system used to check if files are transmitted without errors. Each part of a file goes through the CRC check, and data is re-sent if it fails.

CHAT ROOM: Place on-line where you can "talk" to another person or group of people in real time. Communication is usually through typing. Sometimes they are included as part of a webcam site or simply an on-line community site.

COPY AND PASTE: Procedure in which you copy text on a page into a temporary holding spot called your clipboard, then paste it into another area on the page or into another application. This is usually done by selecting the text or object with your mouse and then either right-clicking your mouse and selecting copy, hitting the Control key and the C key at the same time, or choosing Copy from the Edit menu. Once copied, it is simply a matter of placing the cursor where you would like to move the text or objects and pasting it. This is done again in one of three ways: right-clicking and choosing paste, hitting the Control key and the V key at the same time, or by choosing Paste from the Edit menu.

DIRECTORY: (1) Directory of links to Internet sites. These directories may be searchable. Links to other sites are gathered and sorted by category. Examples of directory sites can be seen in the Appendix of this book. (2) "Folder" of files on a disk on a computer

DOMAIN NAME: Registered address for a web site.

DOMAIN NAME SYSTEM (DNS): System of numbers that are actually used by computers—instead of domain names—to locate a web site. When you type in a URL, such as www.yahoo.com for example, the browser contacts a DNS server, which looks for a DNS address that matches the URL address you entered. To have your own domain, such as www.reallyneatwebcam.com, you would need to have a static IP address, and you also would need to register your domain with a company such as Network Solutions (www.networksolutions.com).

DOWNLOAD: To copy a file from another computer to yours. Technically, this can be anything from accessing a web page to downloading the latest MP3 file of your favorite musical group. The speed at which you download depends on the speed and type of connection to the Internet you have.

DRIVER: Piece of software that allows an operating system to communicate with a piece of hardware. For peripherals and accessories, these are usually provided by the manufacturer of the peripheral. It is important to make sure you have the most current version of a driver before attempting to troubleshoot a problem.

E-COMMERCE: The use of the World Wide Web as a means to do business.

E-MAIL (Electronic Mail): Means of communication over the Internet. No matter how fancy it may look, all e-mail is actually text-based. Attachments of files, such as documents or applications, can be added, but the message itself is text. Even so-called HTML-format e-mail is text-based; however, it is rendered into a web page by your e-mail reader. Be aware that if you send HTML-formatted e-mail, not everyone can read it as such. Also, file attachments such as video e-mail can significantly slow the recipient of a message from receiving it.

FAQ: Frequently Asked Questions. Many web sites try to gather questions that are frequently asked and put them in one place. This is a great first place to start if you have a question.

FIREWALL: Hardware- or software-based system that attempts to prevent unauthorized access to or from a private network. All data (usually broken down into distinct "packets") entering or leaving the intranet passes through the firewall, which can then examine it to see if it meets certain predetermined criteria. Based on that, it will either forward the data on or stop that data transfer.

FTP: File Transfer Protocol. Used to transfer files from one computer to another. It is often used by webmasters to upload their files, such as webcam images, to a web site. It is also used to download files from web sites.

G, GIG, GIGABYTE: Unit of measurement that tells the size of a device's memory or storage capacity. One gigabyte is equal to 1,000 megabytes. Due to its large size, it currently is mostly limited to describing the size of hard disks.

HOME PAGE: Start page that loads whenever you open your browser. This can be changed in your browser's settings.

HYPERTEXT MARKUP LANGUAGE (HTML): Set of codes used to design a web page. Many web page design applications no longer require you to manually code pages; however, efficiency is lost when it is done automatically by the software. Whether or not you initially code the pages yourself, it is a good idea to have some knowledge of HTML and to check the automatically generated page for redundancies.

HYPERTEXT TRANSFER PROTOCOL (HTTP): Industry standard used by all web servers to provide rules for moving text, images, sound, video, and other multimedia files across the Internet.

ICON: Small picture used in graphical interface environments to denote either a function of an application or a link to information or resources.

INTERNET: Worldwide collection of computer networks. All information from web browsers, e-mail clients, newsgroup readers, and file transfers can cross this network.

INTRANET: Localized collection of computer networks, usually used to refer to the networks of a business. Most businesses have a firewall between their intranets and the Internet to protect their users and systems.

IP ADDRESS: Numeric identifier for a computer or device on a TCP/IP network. The Internet and the networks attached to it use the TCP/IP protocol to deliver messages based on the IP address of the destination. IP addressees are expressed or written as four numbers separated by periods. Each number can be a value anywhere from zero to 255. When you always have the same IP address, it is called static; however, most individuals attached to the Internet or any TCP/IP network are given a new IP address when they connect.

ISP (INTERNET SERVICE PROVIDER): Generic term for any company that offers a connection to the Internet. This can be through any number of connection types, including dial-up, cable modems, or DSL.

JAVA: Computer-programming language. Raw Java source code files are compiled into a format called bytecode, which can then be executed by a Java interpreter. Java interpreters and run-time environments, known as Java Virtual Machines (VMs), exist for almost every operating system in use today. Java applets are small applications that can be downloaded from a web server and run on your computer by a Java-compatible web browser, such as Netscape Navigator or Microsoft Internet Explorer.

LAN: Local Area Network. A LAN is used most often to connect computers in a business environment to one another and to a server. Connections over a LAN are generally much faster than connections made over the Internet.

MODEM: Stands for modulate-demodulate. A device that allows computers to communicate with each other over telephone lines or other delivery systems by changing digital signals to analog signals for transmission and then back to digital signals. Modems come in different speeds, and different types of modems are used for different types of connections: ISDN modems, DSL modems, cable modems, and analog-telephone modems.

M, MEG, MEGABYTE: Unit of measurement that tells the size of a device's memory or storage capacity. A standard 3.5-inch disk can store 1.44 MB of data. When used with the term RAM, it means how much memory the computer has available to load parts of programs.

MEGAHERTZ: Usually expressed as MHz. This is a measurement of processing cycles that roughly tells the speed of a computer's processor. The higher the number, the more millions of calculations the processor can achieve each second. One megahertz is equal to one million processor cycles per second.

MOTHERBOARD: Main circuit board of a computer that contains the connectors for attaching additional boards. Usually it is where the CPU, BIOS, memory, mass storage interfaces, serial and parallel ports, expansion slots, and all the controllers required to control standard peripheral devices are found. On most PCs, it is possible to add memory chips directly to the motherboard, which is often needed to increase the computer's performance. In addition, you can increase performance by replacing the CPU chip; however, there is a fair amount of research that needs to be done first. Never assume that a chip from one computer can be used to replace the chip in another.

MULTI-TASK: The ability of a computer to handle multiple operations at once, usually referring to multiple applications. The larger the cache and RAM memory, the better able to multi-task the computer will be. Entry-level processors such as the Intel Celeron have small caches, and thus are not able to handle multiple applications running simultaneously very well.

NETIQUETTE: Rules or manners for interacting courteously with others on-line (such as not typing a message in all capital letters, which is equivalent to shouting).

NEWBIE: Used derisively to describe a person who is new to either computers and the Internet in general, or to a specific aspect of technology. Just remember that if anyone uses it to refer to you that everyone was a newbie at one point in his or her life!

PC CARD: Standard for adding hardware options to notebook computers. Formerly called PCMCIA, PC Card devices usually fit in one of two slots on the side of a notebook, though some do take up two slots. These devices can act as any of a number of traditional PC devices, such as modems, or provide unique mobile capabilities, such as GPS tracking or wireless Internet connectivity.

PLUG-AND-PLAY: Term used to describe the process by which an operating system can detect and properly configure a new device. With some devices this can be done without having to reboot the computer, in which case it is called "hot plug-and-play." While initially problematic under Microsoft Windows, recent versions of Windows 98 and Windows 2000 have greatly improved its effectiveness.

PORT: Interface or connection point on a computer to which you can connect a device or cable. Every PC has various types and numbers of ports. Most computers have one or more serial ports and at least one parallel port for connecting a printer. In addition, you will most likely have a port for your graphics adapter, a network card, and perhaps a joystick. Many manufacturers now color code their ports to ease the confusion, so check your manual as needed.

POST: Act of putting information on the Internet. You usually post to newsgroups or bulletin boards. You can also post a web page.

RAM: Random Access Memory. This is a temporary storage chip or chips that a computer uses to store data in use by programs. Since reading data from RAM is much quicker than reading it from a disk, the more RAM a computer has, the faster it can operate, especially when using complex programs.

REMOVABLE MEDIA: Any type of computer media such as a floppy diskette that can be used to record or access data and then store it away from your computer.

SEARCH ENGINE: Program that performs searches, based on keywords, for information on the Internet. Examples are Yahoo and AltaVista.

SPAM: Expression for unsolicited e-mail, generally from a company trying to get you to visit its web site or an offer to make money. Spam is against most ISPs accepted use policies and can be reported, resulting in offenders losing their account. This is often difficult, since spammers frequently try to hide their account information by providing false data.

TCP/IP: Transmission Control Protocol/Internet Protocol. Suite of protocols on which the entire Internet is primarily based.

URL (UNIFORM RESOURCE LOCATOR): World Wide Web address of a site on the Internet. This is then translated using DNS into an IP address. For example, the URL for the White House is http://www.whitehouse.gov.

UNIVERSAL SERIAL BUS (USB): Latest type of connection for peripherals. It is mainly known for being faster than traditional serial and parallel ports and for allowing hot plug-and-play configuration.

USENET NEWSGROUPS: System of thousands of special-interest groups to which readers can send or "post" messages. These messages are then distributed to other computers on the network. Usenet registers newsgroups, which are available through Internet Service Providers.

VIRUS: Piece of programming code inserted into other programming to cause some unexpected and usually undesirable event, such as lost or damaged files. Viruses can be transmitted by downloading programming from other sites, opening e-mail attachments, or can be present on a diskette or other removable media. Anti-virus programs can protect you from most viruses and are a good investment for almost all computer users.

WORLD WIDE WEB (WEB or WWW): International hypertext-based system that allows you to browse through a variety of linked Internet resources organized by colorful, graphics-oriented home pages. Individual pages are coded in HTML, and then rendered into formatted text and graphics by your browser.

Index

Symbols

A

B

C

D

E

F

G

H

I

J

K

L

M

N

O

P

Q

R

S

T

U

V

W

Y

Z